The Seven C's of Creativity

Series Editor
Todd Lubart
Université Paris Cité
Paris, France

This series explores a new framework for understanding creativity based on seven "C's": Creators (characteristics of creative people), Creating (the process of creating), Collaborations (creativity in dyads, teams, groups), Context (the environment for creativity), Creations (the nature of creative productions), Consumption (the uptake of creative productions, adoption of creative ideas by the public), Curricula (the development and teaching of creativity).

Comprising seven edited volumes, one for each C, the series will be first comprehensive overview of the Seven C's framework and will bring together leading international scholars to study creativity-related phenomena with an interdisciplinary lens. Topics covered will include: group creativity; leadership and creativity in the workplace; consumer creativity; language as a creative consumption medium; writing, screen writing, and filmmaking; creative products in education, music, science, and engineering; creativity and AI; creativity in school and learning; creativity training.

Nathalie Bonnardel
Editor

Collaborations in Creativity

Exploring Human and Human-Computer Interactions

palgrave
macmillan

Editor
Nathalie Bonnardel
Aix Marseille Université, Research center for the Psychology of Cognition, Language and Emotion/Centre de recherche en Psychologie de la Connaissance, du Langage et de l'Emotion (PSYCLE, UR 3273) & Institute for Creativity and Innovations of Aix-Marseille/Institut Créativité et Innovations d'Aix-Marseille (InCIAM)
Aix-en-Provence, France

The Seven C's of Creativity
ISBN 978-3-032-07752-3 ISBN 978-3-032-07753-0 (eBook)
https://doi.org/10.1007/978-3-032-07753-0

Cover credit: japatino

This Palgrave Macmillan imprint is published by the registered company Springer Nature Switzerland AG.
The registered company address is: Gewerbestrasse 11, 6330 Cham, Switzerland

To Emeline, Jonathan, Marie-Julie, and David

Preface

Within the 7C's framework for creativity, as proposed by the Editor of this series, Pr. Todd Lubart, the present volume is dedicated to the "C" of Collaborations. Creativity is rarely a solitary endeavor: more often it emerges through interactions and exchanges between individuals, groups, and increasingly, with technological partners. In a time when both human collectives and human-computer systems are reshaping the landscape of creativity, the exploration of collaborative dynamics becomes crucial.

This book, *Collaborations in Creativity: Exploring Human and Human-Computer Interactions*, brings together a diverse set of perspectives to better understand the multifaceted ways in which creativity and creative processes unfold within teams, organizations, and socio-technical systems. The nine chapters in this book illustrate the richness and complexity of collaborative creativity. They address not only the cognitive and psychosocial foundations of creative activities performed in collective situations, but also methodological approaches for analyzing and assessing collaborative creative processes, the role played by embodied cognition, and the opportunities as well as challenges raised by artificial intelligence and robotics.

The opening chapter lays out the dual perspective of human-human and human-computer collaborations in creative activities, setting the stage for the themes that follow. Subsequent contributions examine, among others, how collaboration can be circumscribed and evaluated in design settings, how creativity can be understood at collective and organizational levels, and how coding schemes may allow researchers to capture the unfolding of collective creative behaviors and processes. Other chapters explore the multi-level nature of scientific collaboration, the insights drawn from embodied cognition, and the use of historical datasets as sources of mutual inspiration. The last part of the book focuses on the rapidly evolving frontier of creative collaborations between humans, generative artificial intelligence, and social robots, examining theoretical concepts, practical manifestations, and their socio-technical implications

Together, these contributions emphasize that collaborations in creativity are not merely the sum of individual contributions. They are dynamic processes through which interactions, contexts, and tools jointly shape new possibilities. Today, as digital technologies and AI systems increasingly enter our creative ecosystems, it becomes essential to rethink collaboration not only as a human-centered endeavor but also as a hybrid one, where human and non-human actors jointly participate in the creative activity.

I wish to thank all the authors for their thoughtful and stimulating contributions, as well as Todd Lubart, the series editor, whose vision for the 7C's provides the unifying framework for this collection. It is my hope that this book will inspire researchers, practitioners, and students to consider collaboration—human and human-computer alike—as a fertile ground for both understanding and fostering creativity.

Aix-en-Provence, France
August 20, 2025

Nathalie Bonnardel

Contents

1 **Human–Human and Human–System Collaborations Towards Creative Activities** 1
Nathalie Bonnardel, Fabien Girandola, Agnès Lellouche-Gounon, Cynthia Lopez-Bagousse, and Karima Toumi

2 **Circumscribing and Appraising Collaboration in Design** 45
Michael J. Baker, Jean-Marie Burkhardt, Françoise Détienne, and Stéphane Safin

3 **The Psychosocial and Cognitive Underpinnings of Collective and Organizational Creativity** 75
Fabien Girandola and Nathalie Bonnardel

4 **Social Process Assessment for Recognizing Creativity: A Creative Behaviors and Processes Coding Scheme** 127
Kyle Christenson, Payge Japp, Apurva Patel, Roni Reiter-Palmon, Matthew Cronin, Joseph A. Allen, Marissa Shuffler, and Joshua D. Summers

5 **Scientific Collaboration as Multilevel Interaction Processes: Dialectic Knowledge Creation Across Teams, Communities, and Society** 155
Yasunori Yamanouchi, Sachiko Kiyokawa, and Takeshi Okada

6 **The Story Lab: What Can We Learn from Embodied Cognition?** 189
Samira Bourgeois-Bougrine and Antoine Le Bos

7 **Using a Large Volume of Local Historical Statistical Tables as a Source of Mutual Inspiration and Collective Creativity** 213
Kumiyo Nakakoji, Yasuhiro Yamamoto, and Hibiki Sasaki

8 **Human-AI and Social Robots Creative Collaboration** 245
Sudapa Chompunuch, Florent Vinchon, and Todd Lubart

9 **From Social Encounters to Sociotechnical Creative Collaboration: Perceiving and Engaging with Robots** 275
Nathalie Bonnardel, Maejda Marcon, Ladislav Moták, and Clément Belletier

Index 299

Notes on Contributors

Joseph A. Allen, Ph.D., Professor of Industrial and Organizational Psychology at the University of Utah, Director of the Center for Meeting Effectiveness, University of Utah, Salt Lake City, USA. joseph.a.allen@utah.edu

Michael J. Baker, Ph.D., Research Director CNRS at Télécom Paris, Institut Interdisciplinaire de l'Innovation (i3) - UMR 9217, Paris, France. michael.baker@telecom-paris.fr

Clément Belletier, Ph.D., Associate Professor/HDR in Psychology at the, Université Clermont Auvergne, Laboratoire de Psychologie Sociale et Cognitive (LAPSCO), CNRS, Clermont-Ferrand, France, and member of the Institut Universitaire de France (IUF). clement.belletier@uca.fr

Nathalie Bonnardel, Ph.D., Full Professor in Cognitive Psychology and Ergonomics at Aix-Marseille Université (amU), Founder and Director of the Institute for Creativity and Innovations of Aix-Marseille/Institut Créativité et Innovations d'Aix-Marseille (InCIAM), member of the Research Center for the Psychology of Cognition, Language and Emotion/Centre de recherche en Psychologie de la Connaissance, du Langage et de l'Emotion (PSYCLE, UR 3273), member of the Board of Directors of the International Society for the Study of Creativity and Innovation (ISSCI), and former member of the Institut Universitaire de France (IUF). Responsible for the master's degree in psychology and ergonomics: human factors and information system engineering as well as for the post-master's degree in creativity and innovation of amU, Aix-en-Provence, France. nathalie.bonnardel@univ-amu.fr

Samira Bourgeois-Bougrine, Ph.D., Associate Professor/HDR in Human Factors, Neuroergonomics and Psychology, Université Paris Cité and Univ Gustave Eiffel, LaPEA, Boulogne-Billancourt, France. sbb9799@gmail.com

Jean-Marie Burkhardt, Ph.D., Research Director, Université Gustave Eiffel and Université Paris Cité, LaPEA, Versailles, France. jean-marie.burkhardt@univ-eiffel.fr

Sudapa Chompunuch, Ph.D., Researcher, Bangkok University, The Institute for Knowledge and Innovation Southeast Asia (IKI-SEA), Bangkok, Thailand. sudapa.chom@gmail.com

Kyle Christenson, Ph.D. student in Industrial and Organizational Psychology at Clemson University, Clemson, SC, USA. kmc9@g.clemson.edu

Matthew Cronin, Ph.D., Professor, George Mason University, Department of Management, Fairfax, VA, USA. mcronin@gmu.edu

Françoise Détienne, Ph.D., Research Director Emerita CNRS at Télécom Paris, Institut Interdisciplinaire de l'Innovation (i3) - UMR 9217, Paris, France. francoise.detienne@telecom-paris.fr

Fabien Girandola, Ph.D., Full Professor in Social Psychology at Aix-Marseille Université (amU), Director of the Laboratory of Social Psychology/Laboratoire de Psychologie Sociale (LPS, UR 849), Deputy Manager of the Institute for Creativity and Innovations of Aix-Marseille (InCIAM), responsible for the postgraduate degree in creativity and innovation of amU, Aix-en-Provence, France. fabien.girandola@univ-amu.fr

Payge Japp, Ph.D. student in Industrial/Organizational Psychology, University of Nebraska at Omaha, Department of Industrial and Organizational Psychology, and Center for Collaboration Science, Omaha, NE, USA. pjapp@unomaha.edu

Sachiko Kiyokawa, Ph.D., Associate Professor at The University of Tokyo, Department of Educational Psychology, Graduate School of Education, Bunkyo-ku, Japan. kiyo@p.u-tokyo.ac.jp

Antoine Le Bos, Founder and Artistic Deputy Manager of the Groupe Ouest, Plounéour-Brignogan, France. antoine.lebos@legroupeouest.com

Agnès Lellouche-Gounon, Project Manager of the Institute for Creativity and Innovations of Aix-Marseille (InCIAM), Ph.D. student in Cognitive Psychology and Ergonomics at Aix-Marseille Université (amU), PSYCLE (UR 3273) & InCIAM, Aix-en-Provence, France. agnes.lellouche-gounon@univ-amu.fr

Cynthia Lopez-Bagousse, Ph.D., Research Engineer in charge of the CreaLab (Center of expertise, advice and consultation on creativity and innovation) in the Institute for Creativity and Innovations of Aix-Marseille (InCIAM), Aix-Marseille Université (amU), PSYCLE (UR 3273) & InCIAM, Aix-en-Provence, France. cynthia.bagousse@univ-amu.fr

Todd Lubart, Ph.D., Full Professor in Differential Psychology, Université Paris Cité and Univ Gustave Eiffel, LaPEA, Boulogne-Billancourt, France. President of the International Society for the Study of Creativity and Innovation (ISSCI), Head of the Research Chair "Homo Creativus" (Foundation University Paris Cité). todd.lubart@u-paris.fr

Maejda Marcon, Ph.D. student in Cognitive Psychology and Ergonomics at Aix-Marseille Université (amU), PSYCLE (UR 3273), InCIAM, and LAPSCO (UMR 6024), France. maejda.marcon@univ-amu.fr

Ladislav Moták, Ph.D., Assistant Professor in Psychology at the Université Clermont Auvergne, Laboratoire de Psychologie Sociale et Cognitive (LAPSCO), CNRS, Clermont-Ferrand, France. ladislav.motak@uca.fr

Kumiyo Nakakoji, Ph.D., Full Professor at Future University Hakodate, Department of Media Architecture, Hokkaido, Japan. kumiyo@acm.org

Takeshi Okada, Ph.D., Full Professor at the University of Tokyo, Department of Educational Psychology, Graduate School of Education, Bunkyo-ku, Japan. okadatak@p.u-tokyo.ac.jp

Apurva Patel, Ph.D., Assistant Professor of Mechanical Engineering at Florida Polytechnic University, Lakeland, FL, USA. apurvapatel@floridapoly.edu

Roni Reiter-Palmon, Ph.D., Distinguished Professor of Industrial and Organizational Psychology at the University of Nebraska at Omaha, Director of the I-O Psychology Graduate Program, Director of Innovation, Center for Collaboration Science, Omaha, NE, USA. rreiter-palmon@unomaha.edu

Stéphane Safin, Ph.D., Assistant Professor at Télécom Paris, Institut Interdisciplinaire de l'Innovation (i3) - UMR 9217, Paris, France, and Université Libre de Bruxelles, Faculty of Psychological Science and Education, Research Center for Work and Consumer Psychology, Brussels, Belgium. stephane.safin@telecom-paris.fr

Hibiki Sasaki, Graduate Student at Future University Hakodate, Graduate School of Systems Information Science, Hokkaido, Japan. hsasaki.res@icloud.com

Marissa Shuffler, Ph.D., Professor of Industrial-Organizational Psychology at Clemson University, Director of the DIGITAL Research Lab (Deriving Innovative & riGorous scIence for Teaming And Leading), Clemson, SC, USA. mshuffl@clemson.edu

Joshua D. Summers, Ph.D., Professor of Mechanical Engineering at the University of Texas at Dallas, Associate Dean for the Erik Jonsson School of Engineering and Computer Science, Richardson, TX, USA. joshua.summers@utdallas.edu

Karima Toumi, Ph.D., Aix-Marseille Université, PSYCLE (UR 3273), LPS (UR 849) & InCIAM, Aix-en-Provence, France. karima.toumi@univ-amu.fr

Florent Vinchon, Ph.D., Temporary teaching and research fellow (ATER), Université Paris Cité and Univ Gustave Eiffel, LaPEA, Boulogne-Billancourt, France. florent.vinchon@gmail.com

Yasuhiro Yamamoto, Ph.D., Full Professor at the Future University Hakodate, Department of Media Architecture, Hokkaido, Japan. yxy@acm.org

Yasunori Yamanouchi, Ph.D., Associate Professor at Tohoku University, Institute for Excellence in Higher Education, Sendai, Japan. yasunori.yamanouchi.d2@tohoku.ac.jp

List of Contributors

Joseph A. Allen University of Utah Health, Rocky Mountain Center for Occupational and Environmental Health, 250 E 200 S, Salt Lake City, UT 84111, USA

Michael J. Baker Télécom Paris, Institut Interdisciplinaire de l'Innovation (i3), CNRS UMR 9217, Dépt. SES, 19 place Marguerity Perey, 91120, Palaiseau, France

Clément Belletier Université Clermont Auvergne, Laboratoire de Psychologie Sociale et Cognitive (LAPSCO), CNRS, 17 rue Paul Collomp, 63037, Clermont-Ferrand, France

Université Clermont Auvergne & Institut Universitaire de France (IUF), Clermont-Ferrand, France

Nathalie Bonnardel Aix-Marseille Université, Centre de recherche en Psychologie de la Connaissance, du Langage et de l'Emotion (PSYCLE, UR 3273), 29 avenue Robert Schuman, 13621, Aix-en-Provence, France

Aix-Marseille Université & Institute for Creativity and Innovations of Aix-Marseille (InCIAM), Aix-en-Provence, France

Samira Bourgeois-Bougrine Université Paris Cité and Univ Gustave Eiffel, LaPEA, 71 avenue Edouard Vaillant, 92100, Boulogne-Billancourt, France

Jean-Marie Burkhardt Université Gustave Eiffel and Université Paris Coité, LaPEA, Campus de Versailles, 25 allée des Marronniers, 78008, Versailles, France

Sudapa Chompunuch Bangkok University, The Institute for Knowledge and Innovation Southeast Asia (IKI-SEA), 9/1 Moo 5 Phaholyothin Road, Pathumthani, 12120, Thailand

Kyle Christenson Clemson University, Department of Psychology, 418 Brackett Hall, Clemson, SC, 29630, USA

Matthew Cronin George Mason University, Management Department, Mail Stop 5F5, Fairfax, VA, 22030, USA

Françoise Détienne Télécom Paris, Institut Interdisciplinaire de l'Innovation (i3), CNRS UMR 9217, Dépt. SES, 19 place Marguerity Perey, 91120, Palaiseau, France

Fabien Girandola Aix-Marseille Université, Laboratoire de Psychologie Sociale (LPS, UR 849), 29 avenue Robert Schuman, 13621, Aix-en-Provence, France

Aix-Marseille Université & Institute for Creativity and Innovations of Aix-Marseille (InCIAM), Aix-en-Provence, France

Payge Japp University of Nebraska at Omaha, 6001 Dodge St, Omaha, NE, 68182, USA

Sachiko Kiyokawa The University of Tokyo, Graduate School of Education, Department of Educational Psychology, 7-3-1 Hongo, Bunkyō-ku, 113-0033, Tokyo, Japan

Antoine Le Bos Le Groupe Ouest, La Gare, 29890, Plounéour-Brignogan, France

Agnès Lellouche-Gounon Aix-Marseille Université & Centre de recherche en Psychologie de la Connaissance, du Langage et de l'Emotion (PSYCLE, UR 3273), 29 avenue Robert Schuman, 13621, Aix-en-Provence, France

Aix-Marseille Université & Institute for Creativity and Innovations of Aix-Marseille (InCIAM), Aix-en-Provence, France

Cynthia Lopez-Bagousse Aix-Marseille Université, Research Center for the Psychology of Cognition, Language and Emotion (PSYCLE, UR 3273), 29 avenue Robert Schuman, 13621, Aix-en-Provence, France

Aix-Marseille Université & Institute for Creativity and Innovations of Aix-Marseille (InCIAM), Aix-en-Provence, France

Aix-Marseille Université, Apprentissage, Didactique, Évaluation, Formation (ADEF, UR 4671), Aix-en-Provence, France

Todd Lubart Université Paris Cité and Univ Gustave Eiffel, LaPEA, 71 avenue Edouard Vaillant, 92100, Boulogne-Billancourt, France

Maejda Marcon Aix-Marseille Université, Centre de recherche en Psychologie de la Connaissance, du Langage et de l'Emotion (PSYCLE, UR 3273), 29 avenue Robert Schuman, 13621, Aix-en-Provence, France

Aix-Marseille Université & Institute on Creativity and Innovations of Aix-Marseille (InCIAM), Aix-en-Provence, France

Université Clermont Auvergne, Laboratoire de Psychologie Sociale et Cognitive (LAPSCO), CNRS, Clermont-Ferrand, France

Ladislav Moták Université Clermont Auvergne, Laboratoire de Psychologie Sociale et Cognitive (LAPSCO), CNRS, 17 rue Paul Collomp, 63037, Clermont-Ferrand, France

Kumiyo Nakakoji Future University Hakodate, Department of Media Architecture, 116-2 Kamedanakano, Hakodate, 041-8655, Hokkaido, Japan

Takeshi Okada The University of Tokyo, Department of Educational Psychology, Graduate School of Education, 7-3-1 Hongo, Bunkyō-ku, 113-0033, Tokyo, Japan

Apurva Patel Florida Polytechnic University, Department of Mechanical Engineering, 4700 Research Way, BARC 1176, Lakeland, FL, 33805, USA

Roni Reiter-Palmon University of Nebraska at Omaha, Department of Psychology, Dodge St. 6001, Omaha, NE, 68182, USA

Stéphane Safin Télécom Paris, Institut Interdisciplinaire de l'Innovation (i3), CNRS UMR 9217, Palaiseau, France

Université Libre de Bruxelles, Faculty of Psychological Science and Education, Research Center for Work and Consumer Psychology, 50 av. F. Roosevelt, 1050, Brussels, Belgium

Hibiki Sasaki Future University Hakodate, Department of Media Design, 116-2 Kamedanakano, Hakodate, 041-8655, Hokkaido, Japan

Marissa Shuffler Clemson University, Department of Psychology, Clemson, 418 Brackett Hall, SC, 29630, USA

Joshua D. Summers University of Texas at Dallas, Department of Mechanical Engineering, ECSS 2.504, School of Engineering and Computer Science, 800 W. Campbell Dr., Richardson, TX, 75080, USA

Karima Toumi Aix-Marseille Université, Centre de recherche en Psychologie de la Connaissance, du Langage et de l'Emotion (PSYCLE, UR 3273), 29 avenue Robart Schuman, 13621, Aix-en-Provence, France

Aix-Marseille Université & Institute for Creativity and Innovations of Aix-Marseille (InCIAM), Aix-en-Provence, France

Aix-Marseille Université, Laboratoire de Psychologie Sociale (LPS, UR 849), Aix-en-Provence, France

Florent Vinchon Université Paris Cité and Univ Gustave Eiffel, LaPEA, 71 avenue Edouard Vaillant, 92100, Boulogne-Billancourt, France

Yasuhiro Yamamoto Future University Hakodate, Department of Media Architecture, 116-2 Kamedanakano, Hakodate, 041-8655, Hokkaido, Japan

Yasunori Yamanouchi Tohoku University, Institute for Excellence in Higher Education, 41 Kawauchi, Aoba-ku, Sendai, 980-8576, Miyagi, Japan

List of Figures

Fig. 1.1	The elaboration and maintenance of a common frame of reference during collective creative activities (adapted from Bonnardel, 2024)	13
Fig. 1.2	A heuristic overview of the MIP-G model (adapted from Nijstad et al., 2019, p. 92)	15
Fig. 1.3	The CEM model (adapted from van Knippenberg et al., 2004)	16
Fig. 2.1	Collaboration defined in relation to collective activity, coordination and cooperation	49
Fig. 3.1	Conceptual links among creative persons, processes, situations, and products (based on Woodman et al., 1993, p. 309)	78
Fig. 3.2	Hypothesis linkages among factors related to organizational creativity (based on Woodman et al., 1993, p. 311)	79
Fig. 3.3	A model of collective creativity and innovation (based on Paulus & Dzindolet, 2008, p. 230)	80
Fig. 3.4	The cognitive-social-motivational model of group creativity (based on Paulus & Kenworthy, 2019, p. 288; Bonetto & Girandola, 2023, p. 123)	85
Fig. 3.5	A theoretical model of factors facilitating divergent and convergent collaborative creativity (DCIM), including intervening processes linking divergent and convergent creativity (based on Coursey et al., 2019a, 2019b, p. 3)	86
Fig. 3.6	Experimental design (based on Kenworthy et al., 2024, p. 1457)	89

Fig. 3.7 Effects of servant leadership on creativity (based on Wang et al., 2022, p. 4) 91
Fig. 3.8 The Analogies and Constraint Management model (based on Bonnardel, 2000; Bonnardel & Lellouche-Gounon, 2024) 97
Fig. 3.9 Types and sources of constraints (based on Cromwell et al., 2018). Constraints are divided into two dimensions: horizontal (resources on the left, problems on the right) and vertical (internal at the bottom, external at the top). In the center, a dynamic problem-solving process unifies the "problem-first" and "idea-first" models. This model integrates these approaches while allowing for flexible iteration between them 99
Fig. 4.1 The Stoplight Model of Creativity (adapted from Cronin & Loewenstein, 2020) 132
Fig. 5.1 Collaborations at three levels in scientific activities 159
Fig. 6.1 Activities and stages of the writing as well as interactions with others, job context, and demands 196
Fig. 6.2 Populating the GREENHOUSE with images, keywords, drawings, etc. 201
Fig. 6.3 Exploration of the GREENHOUSE from different perspectives 203
Fig. 6.4 The flash-light phase within the "GREENHOUSE" 203
Fig. 6.5 "Entering" the story in the 3D narrative space 205
Fig. 7.1 Hakodate City History: Statistical Archives (published in 1987, covering the period from the 1860s to the 1970s) (Hakodate City, 1987) 214
Fig. 7.2 Collective creativity mediated by external representations 217
Fig. 7.3 A cyclic SLAPUP process. The diagram shows a series of concrete actions in the SLAPUP workflow, complementing the abstract five-stage cycle introduced in the text 226
Fig. 7.4 SUM type 1: solo reading of tables from a physical copy of the book (a digital clock on the table tracks elapsed time since the start of the workshop session. The same type of timer is visible in Figs. 7.5, 7.6, 7.7, and 7.9) 228
Fig. 7.5 SUM type 2: solo reading of a single table on an iPad 229
Fig. 7.6 SUM type 3: group reading of a single table projected on a large screen 230
Fig. 7.7 SUM type 4: group reading of a single table printed in a large format 231
Fig. 7.8 DDrPP Magnets. Left: user-facing device with a round wooden disk, held to perceive statistical data as physical resistance. The display reads "year: 1957," indicating that the sensed value corresponds to a record from that year. Right: the internal mechanism, including magnets and electronic components 232

Fig. 7.9 SUM type 5: solo reading of a single table by using DDrPP 232
Fig. 7.10 Progressive enrichment and deepening of the statistical archive data space 236
Fig. 7.11 An example of recorded annotations (from SUM001A; the original is written in Japanese) 237
Fig. 7.12 A screenshot of Stats.Hakodate Browser 238
Fig. 7.13 Replaying the annotation process for the table 239
Fig. 8.1 The involvement of humans and AI in three modes of human-AI collaboration 249
Fig. 8.2 YOLO Robot (open access at https://www.wevolver.com/article/yolo-robot) 253
Fig. 8.3 Example results from idea evaluation 255
Fig. 8.4 Example results from deploying and implementing idea from Leo AI 257
Fig. 8.5 Shimon Robot (open access at https://www.shimonrobot.com/) 260
Fig. 8.6 Rootport "Cyberpunk: Peach John" Manga series (from Midjourney) 261
Fig. 8.7 AI-art creative process diagram (adapted from Elgammal, 2018) 264
Fig. 8.8 The Robotic artist AARON in Living Color (https://whitney.org/exhibitions/harold-cohen-aaron) 265
Fig. 8.9 AlphaGo, AI board game Go match against Lee Sedol (open access from https://deepmind.google/research/projects/alphago/) 267
Fig. 8.10 Conceptual integration of hybrid intelligence/creativity (from Dellermann et al., 2019) 268

List of Tables

Table 1.1	Metaphors and perspectives associated with colors of De Bono's Thinking Hats	22
Table 2.1	Dimensions of collaboration in relation to appraisal of its quality (Détienne et al., 2012b)	64
Table 3.1	The main factors identified in this chapter that contribute to a sociocognitive environment that stimulates creativity.	106
Table 4.1	Exhaustive List of SPARC Codes and Abbreviations	136
Table 4.2	Examples of "Null-Coded" Sense Units	139
Table 4.3	Basic Descriptions of each SPARC code	140
Table 4.4	Examples of problem construction codes	141
Table 4.5	Examples of idea generation codes	141
Table 4.6	Examples of idea elaboration codes	142
Table 4.7	Examples of idea evaluation codes	143
Table 4.8	Excerpt from Mark's problem	146
Table 4.9	Excerpt from Space Hotel problem	147
Table 7.1	The contents of the Hakodate City History: Statistical Archives	215
Table 7.2	The five SLAPUP meeting (SUM) types	227
Table 7.3	Case study—a summary of 13 SLAPUP meetings (the Page# column indicates on which page the statistical table used in the workshop is listed in the Statistical archives book)	234

CHAPTER 1

Human–Human and Human–System Collaborations Towards Creative Activities

Nathalie Bonnardel, Fabien Girandola, Agnès Lellouche-Gounon, Cynthia Lopez-Bagousse, and Karima Toumi

Introduction

In a society marked by rapid changes, environments that are more and more complex, and under an uncertain future, it is increasingly necessary for individuals to be creative, flexible, and competent, in order to perform complex tasks without preestablished procedures (Bonnardel, 2006b/2012; Bonnardel et al., 2023; Vincent-Lancrin et al., 2019). This

N. Bonnardel (✉) • A. Lellouche-Gounon
Aix-Marseille Université, Centre de recherche en Psychologie de la Connaissance, du Langage et de l'Emotion (PSYCLE, UR 3273), Aix-en-Provence, France

Institute for Creativity and Innovations of Aix-Marseille (InCIAM), Aix-en-Provence, France
e-mail: nathalie.bonnardel@univ-amu.fr; agnes.lellouche-gounon@univ-amu.fr

F. Girandola
Institute for Creativity and Innovations of Aix-Marseille (InCIAM), Aix-en-Provence, France

Aix-Marseille Université, Laboratoire de Psychologie Sociale (LPS, UR 849), Aix-en-Provence, France
e-mail: fabien.girandola@univ-amu.fr

N. Bonnardel (ed.), *Collaborations in Creativity*, The Seven C's of Creativity, https://doi.org/10.1007/978-3-032-07753-0_1

is a challenge, acknowledged by many global educational reference systems. Thus, creativity is considered an essential capability for the twenty-first century, alongside with critical thinking, communication, and collaboration. These four capabilities, often referred to as the "4Cs" (OECD, 2019; Joynes et al., 2019; Scott, 2015), are related to the theme of this book, although particular emphasis is placed on both creativity and collaboration.

Creativity is considered a key resource allowing individuals to imagine and propose ideas or solutions which are both new and adapted to the situation or contextually appropriate. This dual specification is consistent with definitions of creativity, which refer to the ability to produce work that is both new and either efficient or considered as having some value (e.g., Bonnardel, 2002; Isaksen et al., 1993; Lubart, 1994; Lubart et al., 2015; Runco & Jaeger, 2012; Sternberg & Lubart, 1999). For instance, creativity can be observed and studied in design contexts, where individuals involved in the design process (e.g., designers, engineers, ergonomics) have to propose ideas and solutions to conceive objects, products, services or processes that are both new—and possibly unexpected—and adapted to the characteristics and constraints of a given situation (Bonnardel, 2006b/2012, 2012, 2021). According to Bonnardel and Lubart (2019),

C. Lopez-Bagousse
Aix-Marseille Université, Centre de recherche en Psychologie de la Connaissance, du Langage et de l'Emotion (PSYCLE, UR 3273), Aix-en-Provence, France

Institute for Creativity and Innovations of Aix-Marseille (InCIAM), Aix-en-Provence, France

Aix-Marseille Université, Apprentissage, Didactique, Évaluation, Formation (ADEF, UR 4671), Aix-en-Provence, France
e-mail: cynthia.bagousse@univ-amu.fr

K. Toumi
Aix-Marseille Université, Centre de recherche en Psychologie de la Connaissance, du Langage et de l'Emotion (PSYCLE, UR 3273), Aix-en-Provence, France

Institute for Creativity and Innovations of Aix-Marseille (InCIAM), Aix-en-Provence, France

Aix-Marseille Université, Laboratoire de Psychologie Sociale (LPS, UR 849), Aix-en-Provence, France
e-mail: karima.toumi@univ-amu.fr

different degrees of *novelty or originality* can be identified, insofar as novelty and originality may be minimal in relation to what already exists (possibly leading to an "incremental innovation"), or may reflect a more significant or even radical change, which is then perceived as a "break" within the existing situation (possibly leading to a "disruptive innovation"). The characteristic of *adaptation or appropriateness to the situation or context* can be interpreted differently depending on the application field (see Bonnardel & Lubart, 2019). For example, an artistic production may be considered contextually appropriate if it satisfies the audience or art critics, or if it brings relief to an artist—or even a patient—as is the case with "art brut," which is not intended to be produced for others. In a different area, a product considered as "ergonomic" must be conceived and developed in order to be adapted to future users' needs, capabilities, and limitations, as well as to their situations of use (Bonnardel, 2006a, 2006b/2012, 2012). Depending on the definitions and/or contexts, the notion of appropriateness or adaptation to the situation can be replaced by other terms, such as relevance, efficiency, usefulness, effectiveness, feasibility, value, etc. Furthermore, Corazza (2016, 2023) added the adjective of *potential* to both originality and effectiveness or appropriateness. This extension allows for incomplete, unrecognized, or emerging ideas to also be considered creative. In this view, creativity would not just based solely on what is produced, but could take into account how it is developed and perceived over time. Thus, what is considered uncreative today may be groundbreaking tomorrow, and vice versa. Therefore, it allows us to acknowledge creativity in the making, and to focus mainly on the *dynamics and processes* (rather than only on results).

Although creativity is acknowledged as essential for the twenty-first century, the ability to be creative is far from a straightforward process. In this chapter, in line with a cognitive and ergonomic psychology approach (Bonnardel, 2006b/2012, 2012), we consider that individuals engaged in a creative activity have to deal with a "problem," which consists in an "open-ended and ambiguous situation, for which one wants and needs new options and a plan for carrying out a solution successfully" (Treffinger, 1995, p. 304; cited by Bonnardel & Lellouche-Gounon, 2024). Such creative problems can be considered as "ill-structured" or "ill-defined" (Eastman, 1969; Simon, 1973) in the sense that the individuals' mental representations are initially incomplete and imprecise (Bonnardel, 2006b/2012, 2012). As pointed out by Tromp (2022, p. 4) "initial states can consist of problem descriptions that are open to different interpretations, contain seemingly contradictory elements, or are so ill-specified that there might be no apparent problem to solve, as is the case in many artistic domains." Therefore, numerous challenges are associated with creative situations, including problem ambiguity, the absence of ready-made solutions, the need to reinterpret information, to deal with multiple constraints, and possibly looking for a balance between novelty and feasibility or value.

More and more *collective creative situations* are observed in our society, since numerous creative problems or situations require knowledge that is not held by a single person but distributed among several actors. In addition to collective situations that are explicitly aimed at achieving creative outputs, creativity also serves to solve ill-defined problems that threaten the status quo or block progress. According to De Dreu et al. (2024), creativity helps individuals and their groups to survive and thrive by adapting to environmental and social challenges. These authors cite several triggers that spark creativity: (1) environmental threats (e.g., Lopez-Bagousse et al., 2024) related, for instance, to climate change or contagious viruses (e.g., Mercier et al., 2021; Pellegrin et al., 2021); (2) economic pressures and resource scarcity; and (3) competition and social conflict, including anticipation of material or immaterial rewards. These triggers for creativity naturally catalyze (or activate) collective creativity by pushing individuals to form groups where their diverse perspectives combine to address complex challenges exceeding individual capabilities. Consequently, we are witnessing a major paradigm shift in how we view creativity. Long dominated by the myth of the "solitary genius," viewing creativity as based on an individual process, the concept of creativity has broadened to recognize the crucial role played by social interactions as it is developed throughout this book. Indeed, in complex creative situations, social interaction, communication, and collaboration are key elements of creative activities (Barrett et al., 2021).

Collaboration is obviously essential when creative work is to be carried out by a group or team, made up of different specialists or professionals with knowledge in different fields, as it is the case, for example, for designing a new car to face ecological issues, or a new habitat adapted to seniors' cognitive and physical limitations. However, collaboration also occurs when the person who is in charge of reaching creative proposals (or the "creator"), for example, a writer, has to interact with other actors, such as their literary agent or publisher. Thus, whether the creator is engaged in a mainly individual creative activity or in a collective creative activity, collaborations occur through *human–human interactions*. If we refer to the seminal model of *4P* proposed by Rhodes (1961), collaboration is related to both the *process* and the *press* (or the environment). This seminal model allows taking into account the person demonstrating creativity, the creative process, the (external) pressure or press, and the creative product or output. To briefly characterize these 4 levels (see Bonnardel & Lubart, 2019), the *person* refers to the various attributes of the individual who creates (such as their intellectual characteristics, personality, or biographical characteristics), the *process* refers to the chain of actions or events involved in creative work, the *press* or pressure from the external environment can be physical or social in nature (and can have a positive or negative influence on creativity), and the *product* refers to the result of creative work and the outputs. We can also consider the framework proposed by Glaveănu (2013), based on a sociocultural and ecological approach. This framework is known as the *5As* model, which refers to *actor*, *action*, *artifact*, *audience*, and *affordance*. According to Lubart (2017), direct correspondences can be established between this theoretical proposition and that of Rhodes (1961), particularly between the actor and the person, between the action and the process, and between the artifact and the product. The audience and affordance can also be interpreted in relation to the external environment, insofar as the environment provides a venue for exhibiting productions to the public, and certain contexts appear more conducive to creative activities than others. Finally, we highlight strong relationships among collaboration, the act of creating and the context, which correspond to three topics of the *7Cs* theoretical framework of creativity proposed by Lubart (2017), which is deeply analyzed in this book series. This framework comprises seven topics consisting of the *creators*, *creating*, *collaborations* during the creative activity, *contexts* in which creativity is expressed, *creations* or creative productions, *consumption* of creative productions, and the *curricula* or training that promotes creativity. More precisely, creating refers to the creative process, the sequence of steps, actions, or events that account for the progress of creators. Studies conducted on this topic can focus on stages of the macro process of creativity, such as incubation, or on micro processes such as the detailed transition from one idea to another during a sequence of divergent thinking, which may vary depending on the domain (see, for instance, Glaveănu et al., 2013). The relations between creating and collaboration will be examined in detail in this book, particularly in Chap. 4 (by Christenson, Japp, Patel, Reiter-Palmon, Cronin, Allen, Shuffer & Summers), which presents a new coding scheme to allow precise analysis of creative behaviors developed in groups or teams. This kind of analysis could be completed by an analysis of the quality of collaboration, as presented in Chap. 2 (by Baker, Burkhardt, Détienne & Safin). *Contexts* refer to the physical and social world in which creators engage in a creative process. Its relations with collaboration will also be explored, particularly in Chap. 3 (by Girandola & Bonnardel), with a deep presentation of psychosocial and cognitive underpinnings of collective and organizational creativity. Indeed, the environment provides resources and constraints that can facilitate or hinder creative behavior, and it also comprises the "field" concerned by the evaluation and selection of creative productions (Csikszentmihalyi, 2014).

Moreover, in this book, we consider that collaborations in creative activities can occur in a physical environment composed of digital devices or virtual agents that are used by the individuals (or Creators) throughout *human–computer interactions* (e.g., Bonnardel & Zenasni, 2010; Bourgeois-Bougrine et al., 2022; Toumi et al., 2021). It is described in this first chapter and it will be later explored in the case of human–AI and robot collaboration both in Chap. 8 (by Chompunuch, Vinchon & Lubart) and in Chap. 9 (by Bonnardel, Marcon, Moták & Belletier). In addition, several domains of human–human collaborative creative activities are going to be analyzed in this book, going from scientific creativity to real-life professional creativity, and citizen creativity. Indeed, scientific creativity and collaboration will be analyzed as multilevel interaction processes, in Chap. 5 (by Yamanouchi, Kiyokawa & Okada). Then, a focus will be placed on creative screenwriting for filmmakers, which will allow, in Chap. 6 (by Bourgeois-Bougrine & Le Bos), the analysis of stories co-development from the perspective of embodied cognition. In Chap. 7 (by Nakakoji, Yamamoto & Sasaki), the focus will be set on cultural activities involving citizens in a museum, with a description of how historical data, interactively provided to users, can serve as an ongoing source of inspiration for collective creativity.

First of all, in this first chapter of the book, key concepts are characterized, especially regarding human–human collaboration and cooperation (some of these notions will be completed in Chap. 2, more especially in the case of design activities). Then, the difficulties encountered in collective creative situations will be pointed out, and conditions necessary for carrying out collaboration, cooperation, and/or co-creativity will be highlighted through models contributing to explaining their interplay and integration in practice. Finally, we will propose approaches to favor and enhance collective creative activities through (1) methods that can be used during human–human interactions and (2) digital systems based on both human–human and human–computer interactions. The studies that will be presented to illustrate these approaches aim especially to address challenges in the context of prospective ergonomics (Brangier & Robert, 2014; Robert & Brangier, 2012), where designers have to imagine and conceive future products that have not yet been developed. Thus, participants or stakeholders have to detect existing user needs, anticipate future ones, and inject creativity into the design solutions they come up with.

Key Concepts: Cooperation, Collaboration and Co-creativity

While the terms of cooperation, collaboration, and co-creativity are often used interchangeably in everyday discourse, they refer to distinct modes of collective engagement, each with unique characteristics and implications for creative outcomes. Clarifying the distinctions between these concepts and characterizing their underlying processes can contribute to a better understanding of how creativity can emerge in group settings and how teams in charge of creative tasks can be possibly supported. Nevertheless, we argue that there is, within the collective creative activities, a continuum between cooperation, collaboration, and co-creativity, and that these notions are not mutually exclusive. In fact, they often coexist within the same creative project, each playing a role at different stages.

Cooperation in Creative Activities

Cooperation refers to a mode of joint activity in which individuals ***work toward a shared objective by dividing tasks and/or responsibilities*** (see also Baker et al, in Chap. 2). In cooperative settings, participants may work largely independently, with coordination primarily concerned with ensuring that individual contributions align with the overarching or superordinate objective. Although the individuals can exchange relevant information and resources in support of each other's goals, creative interdependence remains minimal. Ideas or solutions are usually not jointly developed in the group or the team – contrary to what is observed in the case of collaboration – and creative output does not necessarily arise from the interaction itself, but from the aggregation of actors or stakeholders' efforts. Nevertheless, cooperation requires having one or more shared goals toward which all participants align their actions but it does not necessarily require individuals to coordinate in real time, as tasks can often be divided and distributed among participants. However, coordination is essential to ensure that individual efforts remain aligned with the overarching objective. Thus, actors having different but complementary knowledge and resources can cooperate to reach a shared superordinate and creative objective. It could be the case of a multidisciplinary team comprising specialists in mechanical engineering, in interface design, and in ergonomics, that has to elaborate and develop separate components of a complex product such as a new kind of car addressing ecological issues. These actors can ensure compatibility in their work without deeply interweaving their creative processes.

Collaboration in Creative Activities

Collaboration implies a more integrated form of interaction, in which individuals not only contribute regarding their field of expertise, but *work together* to produce something new and possibly adapted or contextually appropriate (as previously described) *in pursuit of a shared creative objective.* Collaboration emerges as a specific way in which small groups may work together, cooperate, on specific tasks, in real time. As pointed out by Baker et al. (in Chap. 2), collaboration is a particular way of conducting cooperation, where participants strive not only to achieve a shared task, but also a shared understanding of what they are doing together. This concept goes beyond simple cooperation and coordination based on division of tasks: in situations of collaboration, creativity emerges from the collective rather than from juxtaposed individual contributions. It requires that collaborators adhere to a shared vision of what they are trying to achieve, such as the creative problem they are trying to solve, which contributes to aligning their motivations, actions, and efforts (e.g. Bonnardel, 2024). For example, when co-writing a film script, to be able to collaborate, authors must have a common idea of the story they want to tell; without this shared goal, they risk going in incompatible directions. The collective goal therefore provides a common direction for the group's work and it allows the actors to engage in mutual negotiation, idea refinement, and joint decision-making.

Co-creativity

The concept of co-creativity is both complementary to and distinct from cooperation and collaboration, particularly in its emphasis on *emergent creativity through mutual transformation and reciprocal creative engagement*. It corresponds to a dynamic process in which several actors (or agents) jointly engage in the production of new and efficient or valuable ideas or artifacts, and where *creative agency* is distributed and evolving. It is characterized by real-time mutual responsiveness, shared affective involvement, and it highlights the relational and situated nature of creative processes.

Interplay and Integration in Practice: Characteristics and Difficulties

In many real-world contexts, the modes of cooperation, collaboration, and/or co-creativity are not mutually exclusive, but complementary, and there even could be a continuum between these processes. For instance, a design team may begin by cooperating to divide initial tasks, collaborate to make key design decisions, and then enter into moments of co-creativity, where unexpected solutions emerge through synergistic interactions. The transition between modes is dynamic and influenced by context, group dynamics, and task complexity. Carrying out these processes present specific characteristics in practice, and is associated with difficulties and risks.

Characteristics of Collective Creative Activities in Practice

- In the case of collaboration—and co-creativity—ideas emerge and are developed *jointly* rather than individually, which is clearly visible during exchanges among actors or stakeholders. For example, in a brainstorming workshop, one person may propose a draft concept that others enrich, modify, or transform until they collectively arrive at a satisfied new and appropriate solution. Towards this end, each new proposal from a member is accepted, discussed, or adjusted by the others—as it will be analyzed with the coding scheme presented in Chap. 4 (by Christensson et al.). Therefore, the final idea or creative production is co-developed and co-created step by step by the entire group, and it cannot be attributed to a single person. This approach contrasts with traditional teamwork, where each person carries out their part of the project independently: in co-construction, the boundaries between contributions become blurred in favor of a truly joint creation.

- The dynamics of creative collaboration—and co-creativity—are characterized by a constant *back-and-forth exchange of influences* between actors or stakeholders. Each participant not only contributes their own ideas, but also responds and adapts to those of others. There is a strong interdependence: one person's action or suggestion changes the thinking of another, and vice versa, in a continuous cycle. This mutual adaptation ensures that creative development is fostered by everyone's contributions.
- The final creative production usually exceeds what the participants had initially imagined individually, which can result from the cross-pollination phenomenon described by Osborn (1953). By combining different perspectives, the group manages to generate ideas that are more inventive than the simple sum of individual contributions (Sawyer, 2007). Therefore, the final result can be surprising, even to the participants themselves. This *emerging creativity* is an indicator that the process has been successful: the group has arrived at an idea or output that exceeds the sum of its parts, thereby establishing the unique value of collective creativity.

Difficulties and Risks Observed in Collective Creative Situations

Although benefits of collaboration during collective creativity have been highlighted, such as an enrichment of viewpoints and perspectives, difficulties and/or risks are frequently described in collective creative situations. These difficulties and risks can be related to both cognitive and social aspects (e.g., Nijstad et al., 2003a, 2003b), such as the following:

- Misunderstanding and communication problems between group members.
- Lack of cognitive diversity, when groups tend to produce repetitive or narrow ideas, limiting creative potential (Milliken et al., 2003).
- Cognitive interferences between ideas generated by other participants and one's own ideas generation activity (Nijstad et al., 2003a). It can occur, for instance, when proposals are influenced or even blocked by ideas proposed by other members of the group or when one's ideas are constrained by the conceptual or semantic fields

considered by other members of the group. This can be related to the phenomenon of "fixation" (Botella et al., 2023; Jansson & Smith, 1991; Purcell & Gero, 1996; Ward et al., 2002), for example, when the actor remains within the conceptual field of the previous ideas. Other difficulties can be also observed in developing ideas in the event of rapid changes in conceptual fields.

- Production blocking, when only one person can speak at a time in group settings, which leads to interruptions, forgotten ideas, and reduced cognitive flow (Diehl & Stroebe, 1987).
- Lack of focus on others' ideas, in the case of internal recall of ideas for not forgetting one's own ideas.
- Social conformity when individuals suppress unconventional ideas to align with the group consensus, which reduces originality and favors "safe" solutions (Nemeth & Ormiston, 2007).
- Social laziness or loafing, when group members reduce their effort, relying on others to carry the creative burden, which weakens group performance (Karau & Williams, 1993).
- Apprehension of evaluation and social inhibition, when individuals fear negative evaluation by other members of the group, which would lead to more "common" solutions in groups (Nemeth & Nemeth-Brown, 2003).

Numerous studies (e.g., Bellard & Delobbe, 2023) suggest that individuals, in individual situations, often generate more ideas and more original ideas than groups, especially during the early stages of creative processes. In other words, working alone could be considered as more effective for generating many ideas, exploring unusual or risky solutions, avoiding social inhibition, and thus both cognitive and social pressures than working in groups (Diehl & Stroebe, 1987; Nijstad & Stroebe, 2006). However, such studies in favor of individual situations often give only a partial view because they focus almost exclusively on divergent thinking and idea generation (McMahon et al., 2016). The findings of studies focusing on other stages of the creative process (e.g., idea selection and development) are less clear (Rietzschel, Nijstad, & Stroebe, 2006) and sometimes even show that groups tend to reach more creative performances. Large-scale research (Wu et al., 2019) published in the famous scientific journal *Nature*, and based on more than 65 million scientific publications, patents, and software programs produced between 1954 and 2014, showed that individuals working alone or in very small teams tend to produce new and innovative ideas, while large teams develop existing ideas. Moreover, groups would tend to choose the most feasible and desirable ideas rather than the most original ones (Rietzschel et al., 2010).

To try to better understand how to keep positive effects of collective creative activities while reducing negative effects, we will first present models that point out conditions to be satisfied (section "Models and Conditions to Be Satisfied in Collective Creative Activities") and, then, some methods and digital environments which can be used to favor collective creative activities (section "Methods to Favor Creative Activities Through Human–Human Collaboration").

Models and Conditions to Be Satisfied in Collective Creative Activities

Common Frame of Reference (COFOR) Model

In collective creative situations, the knowledge necessary to complete a complex creative task is often *distributed across different professionals or stakeholders* having specialized fields of expertise. Although examples will be later provided in various domains (see Chaps. 5–7), such activities can be especially observed and analyzed in the case of creative design situations (see Bonnardel, 2024). As a result, even highly experienced individuals cannot carry out the design process alone. Instead, successful design work requires the collaborative and/or cooperative involvement of a team, bringing together diverse actors and fields of expertise. To evoke again a previous example, designing a new vehicle to face ecological issues may involve mechanical and electronic engineers dealing with technical aspects, ergonomists who analyze future users' characteristics and needs, and marketing professionals who provide complementary insights relevant to the design objectives. Such projects therefore require interdisciplinary and often transdisciplinary collaboration (Darbellay, 2024). In the case of collective design activities (Bonnardel, 2024), each team member contributes from their own field of knowledge and professional perspective. One of the key advantages of this collective design approach lies in its ability to generate a broader *spectrum of viewpoints*, which can foster original ideas and innovative outcomes. At the same time, these differences in background and practice can present *challenges to communication, mutual understanding, and the interpretation of each other's ideas or proposals.*

Moreover, while the team is expected to work toward a shared overarching goal, there is a risk that individual stakeholders pursue partially divergent objectives while mental alignment is essential within any collective setting. This entails building both a ***shared mental representation*** of the problem and a temporal coordination of actions. Beyond avoiding drawbacks of group work, such as social inhibition or interference among ideas (see the following section), a central challenge is to develop a ***common frame of reference*** that is mutually understood and adopted by all team members, as pointed out in the COFOR (COmmon Frame Of Reference) model (Hoc, 2001; Hoc & Carlier, 2002).

In line with Bonnardel (2024), we propose to apply this model to better understand how cooperation but also collaboration can be efficiently carried out during collective creative activities. Three levels of cooperation are distinguished in this model (see Fig. 1.1): Cooperation in Action, Cooperation in Planning, and Meta-cooperation. Thus, each level reflects a different dimension of cooperation, from the execution of creative tasks

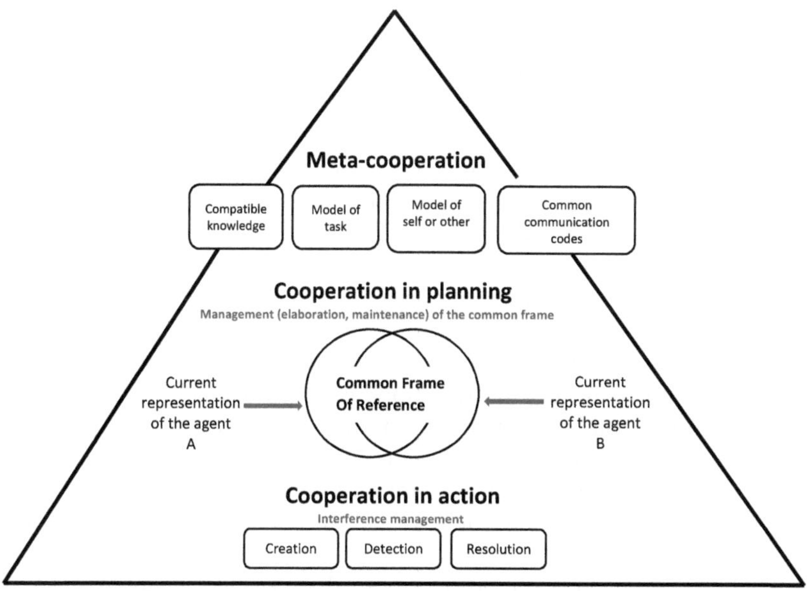

Fig. 1.1 The elaboration and maintenance of a common frame of reference during collective creative activities (adapted from Bonnardel, 2024)

to the coordination of strategies, and meta-reflection on the cooperative process itself. This model accounts for multiple situational parameters:

- At the *meta-cooperation* level, the necessity to have compatible knowledge, to share a common model of the (super-ordinate) task, to have mental representations and perceptions of oneself and of others' tasks, competencies and field of expertise, and to share a communication code;
- At the level of *cooperation in planning*, the actors or stakeholders have to be able to elaborate and maintain a shared representation of the situation or even a common situated awareness in order to align their intentions, goals, and so on;
- At the level of *cooperation in action*, it will then be possible for actors to detect interferences when they are created and to solve them in order to efficiently run the collective creative activity.

Motivated Information-Processing in Groups (MIP-G) Model

As we have seen previously, far from relying on simple individual juxtapositions, groups can become dynamic spaces where cognitive and social diversity is mobilized, transformed, and integrated through sustained and structured exchanges (Nemeth & O'Connor, 2019). The MIP-G model (Nijstad et al., 2019) presents groups as information processors where creativity emerges when individual cognitive resources are effectively developed and shared, and then integrated collectively. This model posits that *motivation* is the channel between available cognitive resources and the effective creative performance of groups. It is structured around two motivational dimensions (see Fig. 1.2):

- *Epistemic motivation*, which corresponds to the desire to analyze information in depth, question assumptions, and systematically explore possible alternatives. In this case, individuals go beyond obvious solutions to seek more original and sophisticated approaches.
- *Social motivation*, which determines whether individuals are oriented toward prosocial goals, such as the collective well-being of the group, or toward goals that favor individual interests. This dimension directly influences the propensity to share knowledge and engage in collaborative development processes.

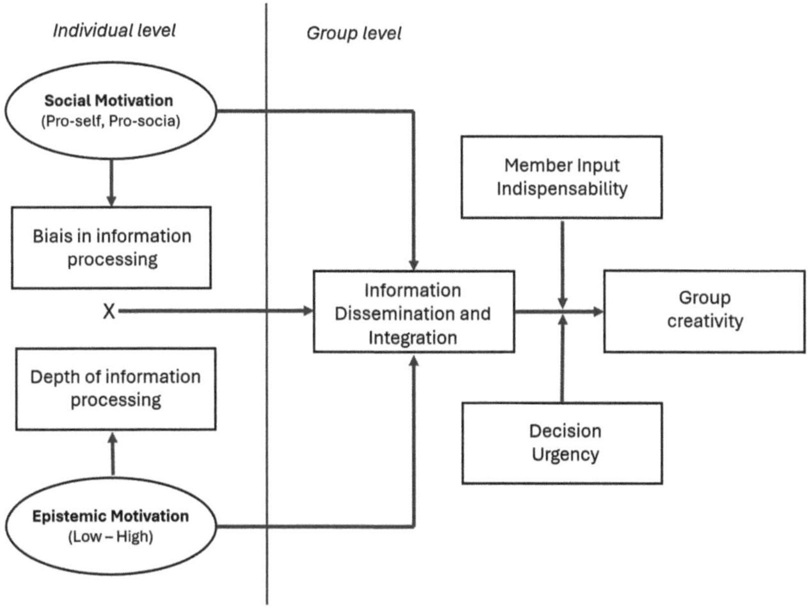

Fig. 1.2 A heuristic overview of the MIP-G model (adapted from Nijstad et al., 2019, p. 92)

According to the MIP-G model, a simultaneous high level of epistemic and prosocial motivation leads to information processing characterized by better knowledge elaboration and integration (De Dreu et al., 2008). Thus, teams combining high epistemic motivation and high prosocial motivation will not only produce more ideas but also more original ideas. However, their effectiveness depends on certain conditions: (1) each member must possess knowledge, skills, or perspectives essential to solving the problem, and (2) low time pressure is a temporal condition because thorough reflection and integration of diverse perspectives require time. In contrast, excessive time pressure would tend to favor quick but uncreative solutions.

Categorization Elaboration Model (CEM)

As highlighted above, *a collective goal* helps the group move forward together and make good use of diverse ideas, avoiding divisions and encouraging creativity. The Social Identity Approach to communication (Greenaway et al., 2014) suggests that social identity and group memberships play an important role in how people process information. This approach draws both on Social Identity Theory (SIT) (Tajfel & Turner, 1979) and Social Categorization Theory (SCT) (e.g., Turner & Oakes, 1986). These two theories assert that individuals internalize group membership as part of their self-identity, and that they evaluate the world around them in accordance with these identities. For example, SIT posits that people exhibit favorability toward information from ingroup sources due to identity reinforcement. Social identity impacts information processing, indicating that individuals may default to skepticism toward information from outsiders or unsourced information. This is also shown through the CEM, developed by van Knippenberg et al. (2004), which provides a better understanding of the effects of diversity on group creativity (see Fig. 1.3). The CEM emphasizes that simply being exposed to ideas is not enough to reap the benefits of diverse information or knowledge. Creative output resulting from collective diversity is largely dependent on categorization performed by members of the team (or stakeholders). If diversity leads to a categorization of group members into "us" *vs* "them (identity threat) this is likely to hinder the sharing of perspectives, resulting in rather low creativity. On the contrary, if it encourages or leads to take into consideration others' contributions, then creativity will be stronger as well as processes involved in creative activities,

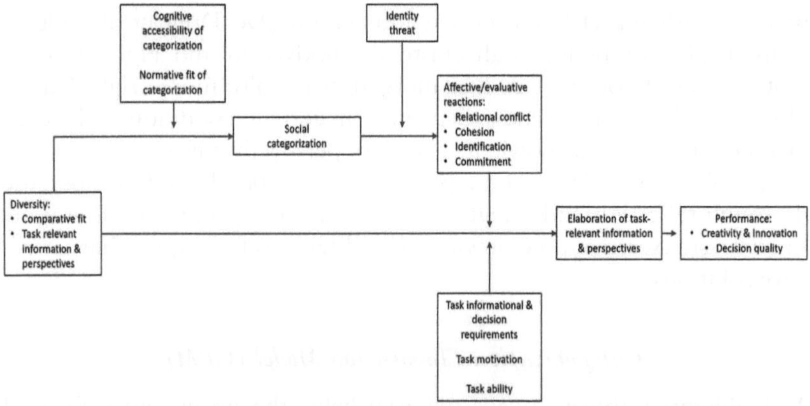

Fig. 1.3 The CEM model (adapted from van Knippenberg et al., 2004)

such as problem solving and effective decision making (Lü et al., 2017; Spoelma & Ellis, 2017; Yang & Konrad, 2010).

The interaction between categorization and elaboration can, therefore, depending on the context, promote or inhibit creativity (e.g., Chen et al., 2019; van Knippenberg et al., 2024). With regard to the CEM, organizations should unlock the innovative potential of their teams by limiting the negative effects of social categorization and promoting elaboration.

PIECE Theoretical Framework

A traditional view considers social norms to be constraints that hinder individual creativity. A norm is an informal social expectation that guides behavior through collective pressure and voluntary compliance. It refers to behaviors or beliefs perceived as acceptable or typical within a group and often carry an implicit expectation to conform (e.g., Cialdini & Trost, 1998; Murrar & Brauer, 2024; Paluck & Shah, 2025). In contrast, a rule is a formal and binding provision imposed by an official authority and accompanied by defined sanctions for noncompliance (see Dubois, 2003). Research shows that certain norms can become levers for activating creative behavior. This is the case with the theoretical framework known as PIECE, proposed by Goncalo et al. (2019). This framework is structured around five key behaviors necessary for collective creativity: (1) **P**articipation in ideation processes, (2) **I**ndependence of thought and creative autonomy, (3) **E**laboration or **e**xploration of alternative solutions, (4) **C**ommunication and **c**ollaborative building on others' ideas, (5) **E**xploration.

Together, these recommended behaviors foster an environment where collective creativity thrives:

(1) One way to boost participation is by encouraging a norm of accountability, the shared expectation of team members that they will be called on to justify their actions and beliefs to others.
(2) Independence is brought about by respecting individual thought and saliency of individualistic norms (Goncalo & Staw, 2006).
(3) The norm of elaboration can be enacted in different ways: some groups put one idea forward then build on it until it is fully developed; others choose to brainstorm many ideas, then evaluate and expand on each of these ideas. This norm allows the group to challenge ideas, which can stimulate elaboration (Postmes et al., 2001).

(4) Communication norms may guide not only how much people say but also how they say it. For instance, norms for equal speaking time can help the group recognize relevant task expertise and promote creativity.
(5) Exploration, inside and outside the group, can be important for developing an idea until it gives rise to an innovation (i.e., norms of risk taking, Dewett, 2007).

Research on these aspects (Goncalo et al., 2019) showed that such principles are effective in stimulating the creativity of naturally less creative members through explicit pressure to conform to pro-creative content. Conversely, highly creative personalities may react negatively to such principles through a psychological reactance mechanism. This PIECE approach can also be illustrated in the case of the arrival of a new member, which can constitute a disruption in the team or group, as discussed in the following paragraph.

Effects of Dissent in Creative Teams

Levine et al. (2019) showed that work groups often perform below their potential because they do not know how to respond creatively when disruptions (Asch, 1956; Diehl & Stroebe, 1987; Steiner's "process loss effect," 1972). For example, the arrival of a new member in the group is an event that can disrupt the cognitive and relational routines of the group (Camacho & Paulus, 1995; Choi & Thompson, 2005). More precisely, the presence of a new member can have two types of influence:

(1) An *involuntary influence* can occur, even before speaking, when the presence of a new member forces the group members to clarify their norms and monitor the distribution of roles. These new developments can increase reflection on procedures, encourage information sharing, and open up space for creative ideas to emerge. The benefits are greater when the group is willing to question its routines and ways of thinking, rather than rigidifying its functioning in order to integrate the newcomer.
(2) An *intentional influence* also occurs when the newcomer seeks to change the group and how it functions, which gives rise to an active minority. If the group members believe that the new comments and suggestions contribute to the collective success, they will be

inclined to consider them. If they perceive them as an obstacle, calling their identity into question, they will ignore or reject them. These factors explain why some teams welcome a newcomer as a lever for change, while others assimilate or expel them without learning anything (De Dreu & West, 2001).

Nemeth and O'Connor (2019) point out that dissent—the persistent expression of a minority viewpoint—acts as a solution to inertia. By breaking unanimity, it triggers a cognitive conflict that forces the majority to process the message in depth, activate broader categories of concepts, and ultimately consider original and creative solutions, even if the dissenter is wrong. In this sense, studies show that encouraging *critical debate* during brainstorming generates more innovative ideas than the classic reminder of the rule against criticism (Nemeth et al., 2004). Three conditions emerge to ensure this dissent or active minority:

(1) Authenticity: a dissident perceived as sincere gains credibility and stimulates cognitive exploration.
(2) Temporal consistency: the stability of a position over several rounds of speech signals conviction and competence, thereby increasing content processing.
(3) Psychological safety: participatory leadership and a norm of open discussion limit fear of punishment and encourage divergent speech, transforming conflict into a learning opportunity (De Dreu & West, 2001).

Thus, according to Nemeth and O'Connor (2019), the goal is not only to remove barriers to creativity (e.g., production blocking, apprehension of evaluation) but to activate social conflict and, beyond that, high-quality cognitive conflict. When the group institutionalized the freedom to disagree, it unleashes broader and more independent thinking, but paradoxically, more collective thinking.

4. Methods to Favor Creative Activities Through Human–Human Collaboration

One of the key points in collaboration during creative activities is communication. If collaboration in creativity leads to the confrontation and integration of diverse perspectives, knowledge, and experiences, it is because dialogue, negotiation, and co-development of new directions take place. It is therefore difficult to separate collaboration from communication; they are closely linked (Thornhill-Miller et al., 2023). These two soft skills are based on cognitive, social, and behavioral competencies, which can facilitate or, conversely, hinder the quality of collective interactions. The performance of a collaboration does not depend solely on individual skills, but, as we have seen in the previous models, on how they are mobilized by the group. For collaboration to be effective, several factors must be in place, such as clear structuring of information, stimulation of social motivations, and a dynamic environment conducive to active participation by everyone (Stasser & Abele, 2020). In the context of collective creative activities, this link between collaboration and communication becomes crucial, as it is based on an explicit goal of problem solving and decision making. Communication supports collaboration by taking the form of exchanges aimed not only at sharing common visions, combining ideas, or clarifying intuitions, but also at verbalizing disagreements, opinions, and cultural or generational biases. Communication should therefore be understood not only as the formulation, presentation, and sharing of ideas with others, but also as active listening and constructive contribution by individuals to collaborative projects, which are often interdisciplinary (Darbellay, 2024). Communication is therefore not simply a tool for collaboration; it must reflect a certain awareness of the collaborative context to play its central role in collective engagement. In collaborative approaches, communication should not be thought of as a simple channel of transmission, but as a fundamental social practice that builds relationships between actors. Therefore, communication is intertwined with collaboration, and together they provide a space for co-creation: a common language is developed and it can be structured with regard to specific rules, as it will be described in the case of the method of De Bono's Six Thinking Hats (section "Studies on the Use of De Bono's Six Thinking Hats in Creative and Collaborative Activities"); meanings and objectives can be further explored and negotiated by looking ahead to the future, as it is the case in the Design-fiction approach (section "Studies on the Use of Design-Fiction in Creative and Collaborative Activities").

Studies on the Use of De Bono's Six Thinking Hats in Creative and Collaborative Activities

In a context of collective human–human creativity, to pragmatically support individuals involved in a design process, we used a creativity method in line with the concepts of the COFOR and MIP-G models, namely the *6 Thinking Hats method* (de Bono, 1970, 1985). To briefly characterize this method in comparison to the seminal brainstorming method proposed by Osborn (1953; see also Chap. 3 by Girandola & Bonnardel in this book), this latter method emphasizes the associative development of ideas while De Bono's method, based on the concept of *lateral thinking* (inspired by the rhizomatic growth of plants), encourages more specifically the exploration of alternative pathways.

In practical terms, when De Bono's Six Thinking Hats method is implemented, participants take turns wearing a hat of a certain color that directs their thinking toward a specific goal. Using De Bono's hats is an individual guidance exercise for collective benefits in small groups (maximum 6 individuals), in which participants agree to play a role ("as if," embodied simulation) as investigators guided by the metaphors of the six colored hats. Each hat color is associated with a specific perspective, a path to explore, in a problem-solving process dynamic (De Bono, 1970, 1985).

In our studies (e.g., Lellouche-Gounon et al., 2023), we chose to ask all participants in workshops using De Bono's hats to wear the different hats in succession, each associated with specific instructions. Having to wear the same color hat at the same time (in line with De Bono, 1991), led them all to go through all six hat colors as they progressed in the collective creative activity (see Table 1.1 for the hats' processing walkthrough). More precisely, a session started with the blue hat, which allowed participants to come up with an initial definition of the problem to be addressed, as well as information about the methodological environment for the task (instructions, resources, questionnaires, ethics). Changes in the hat colors encouraged an evolution in collective thinking and actions, in line with a general ergonomic approach going from the design problem analysis and reformulation or redefinition of the problem, to stages of generation and evaluation of ideas, until they are presented in the form of sketches and plans.

Table 1.1 Metaphors and perspectives associated with colors of De Bono's Thinking Hats

Hat color	*Focus of hats toward a goal and hat metaphor*
Blue 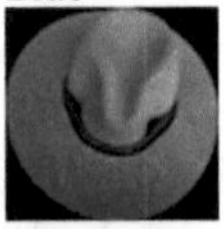	This hat refers to the color of the sky and is the one that allows you to take a step back. At this stage, it is the hat of the big picture, guidelines, organization, and problem framing
Red 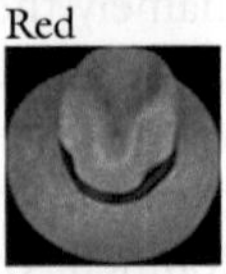	This hat refers to the color of fire and represents emotions. Under this hat, participants are invited to express their feelings "without thinking too much." The aim is to raise awareness of the workshop environment, the roles of the participants, and the desirable and undesirable elements of the problem to be resolved
White 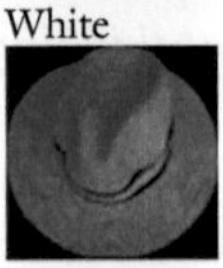	This hat refers to a "neutral" color (e.g., that of printer paper) and represents objectivity, facts, and information
Green 	This hat refers to the color of growing grass and represents unbridled creativity and budding ideas
Yellow 	This hat refers to the color of the sun and represents optimism. Under this hat, participants are invited to detail the advantages and opportunities of each element of the proposed design
Black 	This hat refers to the color of the black sky, risk, and damage and represents pessimism. Under this hat, participants are invited to detail the risks and dangers of each element of the proposed design
Blue again 	Under this hat, participants are invited to detail the next steps and follow-up actions

The process of using the hats aims to increase the participants' perception of the information available in the environment during creative design workshops. The simultaneous transition from one hat to another for all participants constitutes a lever for synchronizing their actions and thoughts on the one hand, and, on the other hand, an opportunity to guide and scaffold information processing and categorization in the form of ideas and sketching (Suwa et al., 1998a, 1998b).

Depending on the studies, the creative tasks proposed in a collaborative setting entailed designing either an ideal connected appartment (consisting in, what is called in France an appartement of "T1" type, which refers to a dwelling with one habitable room with a surface area of approximately 30 m^2), or a bus, for elderly people and their caregivers—whether family or professional. The appartment was designed collaboratively by teams comprising a doctor, family and professional caregivers, and seniors. The bus was designed by master's students in cognitive and ergonomic psychology. Depending on the conditions, the participants in the workshops had access to De Bono's thinking hats and/or personas constructed to reflect the characteristics of future users.

Within each workshop, creativity was first measured with regard to several indicators (using Visual Analog Scales, Delignières, 2016). It showed that the use of De Bono's hats promotes creativity in terms of fluidity, flexibility (number of different solutions), degree of elaboration, usefulness of the solutions considered, consideration of the physical context, and respect for different types of constraints (prescribed, constructed, and deduced—Bonnardel, 1992; see also Chap. 3 in this book).

Qualitative analyses of the quality of collaboration were also conducted in certain workshops, using the "Quality of Collaboration" (QC) analysis grid (Safin et al., 2021; Baker et al., in the following chapter of this book). This qualitative analysis method involves one or more judges analyzing a (filmed) design meeting according to several dimensions associated with indicators such as fluidity of collaboration, mutual understanding, information exchange, argumentation and consensus, task and time management, and cooperative orientation. Here again, the results show a higher quality of collaboration under conditions with De Bono's hats compared to other conditions.

Although the pitfalls of design fixation and groupthink sometimes required the facilitator to intervene to restart the ideation process or its development, the benefits of the lateral thinking approach induced by the six thinking hats method are observed, at least, at two levels:

(1) On the one hand, the rule that participants think simultaneously while wearing the same hat creates a synchronization effect in a

"safe" environment where participants express their opinions under the cover of a hat, in a role-playing exercise that aims to remove inhibitions and stimulates intrinsic motivation to participate actively. Each time the hats are changed, each participant individually and collectively draws on their cognitive resources to adapt them to the challenge of addressing the issue. In our studies, the issues were related to societal challenges (housing and transportation), which lead participants to develop a sense of in-group citizenship and committed social motivation. The information requested from participants was categorized and returned by the participants, which had a positive effect on the development of a "common frame of reference" for the project, in line with the COFOR model. Moreover, each participant, whatever their profile (elderly person, informal caregiver, professional caregiver) expressed their needs and the ones of the future users in a common space, and considered their shared or individual constraints (since individual goals can differ from the common project objective) before proposing ideas. A fluidity was also observed in the workshop phasing, which evolved through cooperation, then collaboration, and finally co-creativity in the activity.

(2) On the other hand, in terms of reducing the negative effects on their activity, such as reducing the impact of cognitive overload during the workshop, participants were aware of ad hoc knowledge, skills, and perspectives and it enabled them to provide best practices consolidated in their long-term procedural memory. These assets enabled them to go further by imagining devices that are better suited than those they already use.

Studies on the Use of Design-Fiction in Creative and Collaborative Activities

In the context of creative activities conducted in group settings, ***design-fiction*** emerges as a stimulating practice. Drawing on science fiction imaginaries, it proposes speculative scenarios, immersive narratives, and narrative artifacts that give shape to alternative futures—sometimes absurd, unsettling, or inspiring—in order to interrogate current sociotechnical dynamics more effectively. This approach does not seek to predict the future, but rather to explore its possibilities by making latent hypotheses or tensions tangible. In doing so, design-fiction opens up spaces for collective imagination and critical reflection, encouraging active engagement with technological, social, or environmental transformations (Lopez-Bagousse et al., 2024). Through its narrative, affective, and participatory dimensions, it contributes to the goals of collaboration during creative activities by providing frameworks for discussion and co-imagination, where users become active participants in envisioning futures.

In the context of an experimental study on future modes of communication, Lopez-Bagousse et al. (2025) analyzed the effects of fiction on dialogic activity and creativity in group settings. Specifically, participants were assigned a collaborative foresight design task that varied in levels of fictionalization, with three experimental conditions: (1) a "control" condition with no fiction; (2) a "foresight" condition, based on the extrapolation of existing trends; and (3) a "design-fiction" condition, grounded in science fiction imaginaries.

Eighteen students in psychology and ergonomics, divided into six groups of three, were asked to collaboratively imagine new uses, objects, or forms of communication in a future context. Their outputs were recorded via audio and video and then transcribed into text to better understand how dialogic activity influences both the problem-solving process and the emergence of creativity within groups. The collected data were analyzed along two complementary axes: one focusing on communication and cognitive mechanisms during dialogue, and the other on the forms of creativity employed.

The first axis required a modeling of dialogic activity to distinguish types of communication based on the nature of reasoning involved. Two types of reasoning were identified: *constructive reasoning*, aimed at generating solutions, and *argumentative reasoning*, intended to present perspectives and raise questions. This analysis highlighted the dual function of communication in collaborative settings: one function fosters collaboration through the sharing of common visions and idea combination; the other facilitates the clarification of intuitions and the verbalization of disagreements, opinions, and generational or cultural biases.

The second axis focused on analyzing creative performance through the lens of the types of creativity mobilized. The objective was to provide a nuanced analysis of the nature of creativity deployed within the groups, based on the framework proposed by Harvey and Berry (2023), who distinguish between three forms of creativity: (1) *maximizing creativity*, centered on highly original ideas with little concern for utility; (2) *balancing creativity*, which seeks to reconcile moderate levels of novelty and usefulness by managing the tension between the two; and (3) *integrative creativity*, where originality and usefulness are coherently combined.

Results showed that the "design fiction" condition, characterized by a high degree of fictionalization, significantly enhanced both argumentative exchanges and creative performance. More precisely, this condition led to outputs that integrate both originality and utility, corresponding to what is termed *integrative creativity*. In contrast, the "control" and "foresight" conditions were more often associated with *maximizing creativity*, emphasizing originality without explicit consideration of feasibility or usefulness.

In conclusion, this study highlights the relevance of design fiction as both an epistemological and methodological lever to stimulate creativity in collective and collaborative contexts. It transforms communication itself into a co-constructive space where interactions are not only about conveying visions of the future but also about co-interpreting and co-inventing shared futures. Fiction acts as a common language, liberating imaginaries, facilitating the confrontation of heterogeneous ideas, and neutralizing power or expertise asymmetries—thereby helping reduce social and identity-based categorizations by shifting the discussion onto a shared fictional and speculative terrain. Indeed, according to the Categorization-Elaboration Model (CEM; van Knippenberg et al., 2004), diversity becomes a driver of collective performance and creativity only when it is accompanied by a genuine process of cognitive elaboration. This process requires, on the one hand, a deep and critical exchange of information, ideas, and perspectives, allowing group members to draw on the richness of their differences. On the other hand, it presupposes active and respectful listening, where individual contributions are considered regardless of the social, disciplinary, or professional background of the contributors. The CEM is based on a shared commitment to co-construct knowledge or solutions by constructively confronting diverse viewpoints in pursuit of a common goal. Design-fiction perfectly illustrates the conditions outlined by the CEM to foster creativity and innovation, moving beyond "us vs. them" logics to transform diversity into innovation and creativity.

Moreover, *integrative creativity*, as defined by Harvey and Berry (2023)—that is, a coherent integration of originality and usefulness—can be seen as the outcome of the simultaneous mobilization of several components from the PIECE model (Goncalo et al., 2019). Specifically, this form of creativity typically relies on: (1) active participation from all group members, ensuring collective engagement in the creative process; (2)

intellectual independence, which promotes the expression of original ideas; (3) in-depth elaboration of ideas through collective work involving confrontation, development, and argumentation; and (4) a culture of exploration, encouraging risk-taking, curiosity, and the group's ability to break away from habitual ways of thinking. In this regard, design fiction creates a fictional design space that is socially framed yet flexible enough to allow coordination between divergent and convergent thinking, thereby enabling the emergence of integrative creativity.

5. Digital Environments to Favor Creative Activities Through Human–Human and Human–Computer Collaboration

While face-to-face interactions are often preferred (Denstadli et al., 2012), we now live in a world where the boundary between digital and physical realms is increasingly blurred. To adapt to this reality, information and communication technologies (ICT) have become integral to daily life, fundamentally transforming how we work, study, and socialize. This is particularly relevant given the growing prevalence of geographically distributed design teams (e.g., Gibson & Gibbs, 2006; Maciver & Malins, 2016). Among these technologies, virtual worlds offer specific affordances such as avatars, co-presence, and immersion that foster creative team interactions (Alahuhta et al., 2014; Alahuhta et al., 2016). In this chapter, we focus specifically on the role of avatars in supporting collaborative creativity. Traditionally used in gaming contexts, avatars (i.e., digital representations of users) are increasingly used outside gaming, including in social media for self-presentation, education, and in professional settings (Bosch-Sijtsema & Sivunen, 2013; Inkpen & Sedlins, 2011; Nowak & Fox, 2018; Ratan et al., 2022; Witt et al., 2016). More precisely, in the sections below, we explore how avatar-based manipulation, ranging from dynamic personas to phenomena such as the Proteus Effect, can be strategically employed to foster creativity and collaboration in virtual environments.

Studies on the Use of "Dynamic Personas" in a Virtual Environment

To favor creativity and collaboration in a virtual environment, Bonnardel and her colleagues (Bonnardel et al., 2016; Bonnardel & Pichot, 2020) proposed and tested an innovative method based on dynamic personas, which constitutes an evolution of the "classical" persona approach. It appears particularly useful in prospective ergonomics contexts, where no explicit user requirements exist, and designers must anticipate future needs while fostering creativity within collaborative and possibly geographically dispersed teams. The "classical" persona method (called here "static persona" method) provides participants or stakeholders with fictional yet realistic user archetypes—developed from field data gathered via interviews, observations and/or questionnaires—which are generally associated with concrete representations including both textual descriptions and photos (e.g., Pruitt & Grudin, 2003). The use of this method aims to promote user-centered thinking as well as enhancing the empathy towards future users and creativity of participants in the design process (Brangier et al., 2011). However, one noted limitation is that it does not actively engage participants in the information-gathering process (Goh et al., 2017). To address this issue, we developed the "dynamic persona" approach (Bonnardel et al., 2016; Bonnardel & Pichot, 2020), which shifts from the traditional static format—where participants passively read a predefined information set based on text and images—to a more interactive format, in which actors, such as designers or other stakeholders, actively seek information through dialogue and interaction with a dynamic persona. Concretely, dynamic personas consist of avatars embedded in a virtual environment who represent archetypes of future users (in our studies, dynamic personas were played by the experimenter, but conversational autonomous agents can also be used) that unfold user information gradually through active exchanges.

Two experimental studies were conducted to investigate the benefits of dynamic personas compared to static personas in both professional and academic contexts. All participants had to perform a same collective creative task aiming at proposing ways of reducing traffic jams and the occurrence of road accidents.

- The first study was exploratory (Bonnardel et al., 2016) and it examined, especially among professional dyads (composed of a designer and an ergonomist), how dynamic versus static personas influenced creative performance, collaboration quality, and empathy. The main findings showed that (1) professional participants interacting with a dynamic persona generated more ideas than those with a static persona (fluency); and (2) dynamic persona groups achieved higher originality scores than static persona groups (idea originality). Moreover, the quality of collaboration within the groups, assessed via the same observation grid as the one previously described (see section "Studies on the Use of De Bono's Six Thinking Hats in Creative and Collaborative Activities"; Burkhardt et al., 2009; Safin et al., 2021; and the following chapter by Baker et al.), showed that quality of collaboration was higher in the dynamic persona condition than the static persona condition. Finally, participants interacting with the dynamic persona reported higher empathy levels toward the end user than those in the static condition.

- The second study (Bonnardel & Pichot, 2020) consisted in a large-scale experimental study with 102 psychology students and it allowed us to further evaluate the effects of dynamic versus static personas on creativity, empathy, and collaboration. The obtained results show a significant improvement in the *quality* of creative ideas (regarding both novelty, feasibility, relevance) for participants using dynamic personas. They also confirmed that participants in the dynamic persona condition reported significantly higher levels of reactive *empathy* (towards the future user represented by the dynamic persona) and that the *quality of collaboration* was significantly higher with a dynamic persona, reflecting more fluid interactions, shared understanding, and effective information exchange. Thus, these findings confirm the potential of dynamic personas to enhance collaborative processes and empathy towards future end-users, which positively influence the quality of creative outputs.

The combined empirical evidence highlights the relevance of dynamic personas as a valuable method and environment for enhancing collaboration, empathy, and the quality of creative solutions in design activities. By fostering active engagement with user representations in virtual environments, this method holds promise for supporting distributed design teams and prospective ergonomic projects. Nevertheless, further research is necessary to optimize its implementation in professional settings and explore technological enhancements, including autonomous agents and immersive virtual environments.

While dynamic personas help designers imagine and take into account needs, capacities, and limits of future users, complementary research shows that the way people see their own avatars can also shape how they think and create. The next section explores how visual features, like avatar similarity and the Proteus Effect, and the use of virtual reality, can influence creativity.

Complementary Studies on Characteristics of Avatars During Remote Collaboration in Creative Situations

A growing body of research suggests that avatars can be strategically designed to support creativity in remote collaboration settings, particularly by manipulating their visual characteristics. One promising approach involves altering the appearance of group members to mitigate common challenges of virtual communication. For instance, Guegan et al. (2017) and Buisine and Guegan (2020) have shown that adjusting social identity cues (e.g., avatar uniforms) enhances group cohesion and a sense of belonging, thereby reducing social loafing (Karau & Hart, 1998; Karau & Williams, 1997), generally observed in online contexts (Aer et al., 2025; Shiue et al., 2010). Similarly, studies indicate that communicating with interlocutors represented by avatars similar to or identical to one's own can facilitate smoother interactions and improve persuasion (Shih et al., 2023). The manipulation of avatar appearance can also significantly improve group decision-making and collective performance. Van Der Land et al. (2015) found that the greatest levels of social attraction and collaboration were achieved when avatars displayed visual similarity among team members and were closely aligned with their users. This suggests that, similar to physical interactions, similarity between social partners, even when represented by avatars, is a strong predictor of trust (Clerke & Heerey, 2021; DeBruine, 2002; Fischer, 2009; Glaeser et al., 2000; Ziegler & Golbeck, 2007).

Beyond group similarity effects, studies show that *user-similar avatars* can enhance creativity. De Rooij et al. (2017) and Marinussen and de Rooij (2019) both found that avatars resembling users improved idea generation in comparison to neutral or stereotypical avatars. This effect appears to be driven by increased identification (Marinussen & de Rooij, 2019), which fosters positive affect and, in turn, enhances mental flexibility and originality (Baas et al., 2008; Dreisbach & Goschke, 2004). In the same line, Park et al. (2019) used facial morphing techniques to create personalized avatars for storytelling and found that children who strongly identified with their avatars demonstrated greater creativity in storytelling tasks. Echoing these findings, Gorisse et al. (2023) suggest that user-similar avatars might enhance creativity by reducing cognitive load and improving focus, likely due to stronger identification and increased motivation (see also Birk et al., 2016). Next to enhancing creativity, self-similar avatars can also enhance task engagement (Zhu & Yi, 2024), which is one of the key factors for successful performance in a creative task (Conti et al., 1995; Du et al., 2016).

In addition, studies on the *Proteus Effect*, which refers to how avatar characteristics influence users' behaviors and self-perception (Yee & Bailenson, 2007), show that avatars can enhance creativity through stereotyped appearances. For instance, Guegan et al. (2016) found that engineering students embodying avatars of inventors generated more creative ideas than those using neutral avatars. However, identification with the avatar appears to be a crucial factor (Guegan et al., 2019; Ratan et al., 2020). Despite the growing interest in multiuser virtual worlds and immersive platforms, a more recent trend has emerged in response to the COVID-19 pandemic: the integration of avatars within videoconferencing platforms. This hybrid approach seeks to combine the accessibility of traditional videoconferencing tools with the benefits associated with avatar embodiment. This growing interest in avatar integration within videoconferencing is particularly relevant for creative tasks, which often suffer from the cognitive and social constraints of conventional video calls (Brucks & Levav, 2022; Kristensson & Norlander, 2003). Unlike face-to-face interactions, videoconferencing heightens self-awareness by displaying one's own image on screen, increasing anxiety, cognitive load, and self-monitoring, all of which impair focus and creative performance (Bailenson, 2021; Carver & Scheier, 1990; de Vasconcelos Filho et al., 2009; He et al., 2022). The "hyper-gaze" effect, where participants feel constantly observed due to gallery views and zoomed-in faces, further contributes to stress and inhibits risk-taking essential for creativity (Fauville et al., 2023; Gilovich et al., 2000; Amabile, 1996). These conditions can also lead to social inhibition due to anticipated group evaluation (Cottrell, 1972; Nemeth & O'Connor, 2019). In response to these pressures, users often turn off their cameras, which reduces social presence and visual cues critical for collaboration, potentially leading to disengagement or "social loafing" (Latané et al., 1979; Castelli & Sarvary, 2021; Lengel & Daft, 1988). This action also removes important nonverbal cues that help manage cognitive load (Yung & Paas, 2015) attention (Schneider et al., 2018), and sustain motivation and engagement in digital settings (Kim & Jo, 2020; Xie et al., 2017). The limitations of videoconferencing platforms, underscore the potential of avatars to act as mediating tools that reduce anxiety, support engagement, and foster creative performance in virtual collaboration.

Finally, another promising environment to support the design process and foster creativity is ***virtual reality (VR)***. While early research was constrained by costly and limited hardware (e.g., heavy, low-resolution headsets; Coburn et al., 2017), recent advances have increased VR's accessibility and opened new creative possibilities. As emphasized by Toumi et al. (2021), VR offers both cognitive (e.g., spatial exploration, perspective changes) and social (e.g., collaboration, shared space) affordances that support creative ideation and problem-solving across various design stages, from early brainstorming to prototyping. Bourgeois-Bougrine et al. (2022) further demonstrate that immersive environments can enhance both individual and team creativity through real-time embodied interaction (e.g., avatar synchronization, spatial cues) and emotional engagement.

Conclusion

In collective contexts, creativity emerges from a combination of cognitive stimulation and social interactions, as discussed in this chapter. Indeed, collective creative activities can involve a potential phasing, including cooperation, collaboration, and co-creativity by and between participants. The dynamics at work within groups require specific methodological support or the use of digital tools capable of sustaining collective engagement, encouraging constructive confrontation, and maintaining a high level of creativity. The COFOR reference framework suggests that it is possible to enable team members to work in sync in order to optimize their perception of the initial situation and its evolution, their mental flexibility, and the quality of their collective interactions. The MIP-G model emphasizes that teams in collective idea generation situations act as motivated information processors, depending on the moderating effect of certain factors. The PIECE framework identifies five core behaviors essential to collective creativity: participation, independence, elaboration, communication, and exploration. The CEM suggests how social diversity within teams can sometimes lead to "us versus them" categorization, thereby limiting collaboration and creativity. Taking these models into account highlights the richness and complexity of collective creativity dynamics, particularly in innovative contexts relying on the use of creativity methods such as De Bono's Six Thinking Hats, design fiction, and/or the use of digital devices that allow avatars to be integrated into collaborative environments, as is the case with the Dynamic Personas method. By mobilizing these models, methods, and/or devices in a complementary manner, it becomes possible to design collective environments that foster creativity within groups. Their articulation contributes to transforming collective creativity into a sustainable and innovative process.

References

Aer, B. E., Lapian, S. J., & Gunawan, E. M. (2025). Social loafing in virtual teams: Analyzing the challenges of remote collaborations. *Kultura: Jurnal Ilmu Hukum, Sosial, dan Humaniora, 3*(1), 367–377.

Alahuhta, P., Nordb, E., Sivunen, A., & Surakka, T. (2014). Fostering team creativity in virtual worlds. *Journal For Virtual Worlds Research, 7*(3), 1–24.

Alahuhta, P., Sivunen, A., & Surakka, T. (2016). Virtual worlds supporting collaborative creativity. In Y. Sivan (Ed.), *Handbook on 3D3C platforms: Progress in IS* (pp. 103–121). Springer.

Amabile, T. M. (1996). *Creativity and innovation in organizations* (Vol. 5). Harvard Business School.

Asch, S. E. (1956). Studies of independence and conformity: I. A minority of one against a unanimous majority. *Psychological Monographs: General and Applied, 70*(9), 1–70.

Baas, M., De Dreu, C. K., & Nijstad, B. A. (2008). A meta-analysis of 25 years of mood-creativity research: Hedonic tone, activation, or regulatory focus? *Psychological Bulletin, 134*(6), 779.

Bailenson, J. N. (2021). Nonverbal overload: A Theoretical argument for the causes of zoom fatigue. *Technology, Mind, and Behavior, 2*(1), 1–6.

Barrett, M. S., Creech, A., & Zhukov, K. (2021). Creative collaboration and collaborative creativity: A systematic literature review. *Frontiers in Psychology, 12*, 713445.

Bellard, E., & Delobbe, N. (2023). The interplay between individuals and teams in producing original work. *The Journal of Creative Behavior, 57*(3), 342–356.

Birk, M. V., Atkins, C., Bowey, J. T., & Mandryk, R. L. (2016, May). Fostering intrinsic motivation through avatar identification in digital games. In *Proceedings of the 2016 CHI conference on human factors in computing systems* (pp. 2982–2995). ACM.

Bonnardel, N. (1992). *Le Rôle de l'Evaluation dans les Activités de Conception.* Doctoral dissertation, Aix-Marseille 1. INRIA.

Bonnardel, N. (2002). Entrée: Créativité. In G. Tiberghien (Ed.), *Dictionnaire des Sciences Cognitives* (pp. 95–97). Armand Colin/VUEF.

Bonnardel, N. (2006a). Entrée: Ergonomie : Communication Homme-Machine. In S. Mesure & P. Savidan (Eds.), *Dictionnaire des Sciences Humaines* (pp. 386–388). PUF.

Bonnardel, N. (2006b/2012). *Créativité et Conception: Approches Cognitives et Ergonomiques.* De Boeck Supérieur.

Bonnardel, N. (2012). Designing future products: What difficulties do designers encounter and how can their creative process be supported? *Work, A Journal of Prevention, Assessment & Rehabilitation, 41*, 5296–5303. https://doi.org/10.3233/WOR-2012-0020-5296

Bonnardel, N. (2021). Créativité et conception. In E. Brangier & G. Vallery (Eds.), *Ergonomie : 150 Notions Clés* (pp. 223–227). Dunod.

Bonnardel, N. (2024). Design activities. In F. Darbellay (Ed.), *Elgar encyclopedia of interdisciplinarity and transdisciplinarity.* Edward Elgar Publishing.

Bonnardel, N., Forens, M., & Lefevre, M. (2016). Enhancing collective creative design: An exploratory study on the influence of static and dynamic personas in a virtual environment. *The Design Journal, 19*(2), 221–235. https://doi.org/10.1080/14606925.2016.1129145

Bonnardel, N., Girandola, F., Bonetto, E., & Lubart, T. (Eds.). (2023). *La Créativité en Situations : Théories et Applications.* Dunod.

Bonnardel, N., & Lellouche-Gounon, A. (2024). Constraints in creativity: Their role and use to reach creative (design) solutions. In C. Tromp, R. J. Sternberg, & D. Ambrose (Eds.), *Constraints in creativity* (pp. 77–99). Brill Publishers.

Bonnardel, N., & Lubart, T. (2019). La créativité : approches et méthodes en psychologie et en ergonomie. *RIMHE - Revue Interdisciplinaire, Management, Homme & Entreprise, 37*, 79–98.

Bonnardel, N., & Pichot, N. (2020). Enhancing collaborative creativity with virtual dynamic personas. *Applied Ergonomics, 82*, 102949. https://doi.org/10.1016/j.apergo.2019.102949

Bonnardel, N., & Zenasni, F. (2010). The impact of technology on creativity in design: An enhancement? *Creativity and Innovation Management, 19*(2), 180–191. https://doi.org/10.1111/j.1467-8691.2010.00560.x

Bosch-Sijtsema, P. M., & Sivunen, A. (2013). Professional virtual worlds supporting computer-mediated communication, collaboration, and learning in geographically distributed contexts. *IEEE Transactions on Professional Communication, 56*(2), 160–175.

Botella, M., Pichot, N., Volle, E., Cassotti, M., Lubart, T., & Bonnardel, N. (2023). Créativité et cognition. In N. Bonnardel, F. Girandola, E. Bonetto, & T. Lubart (Eds.), *La Créativité en Situations : Théories et Applications* (pp. 70–85). Dunod.

Bourgeois-Bougrine, S., Bonnardel, N., Burkhardt, J.-M., Thornhill-Miller, B., Pahlavan, F., Buisine, S., Guegan, J., Pichot, N., & Lubart, T. (2022). Immersive virtual environments' impact on individual and col-lective creativity a review of recent research. *European Psychologist, 27*(3), 237–253. https://doi.org/10.1027/1016-9040/a000481

Brangier, E., Bornet, C., Bastien, J. M. C., Michel, G., & Vivian, R. (2011). Mesure de la capacité des personas à générer des idées dans la conception de projets WEB. *Le Travail Humain, 75*(2), 121–145.

Brangier, E., & Robert, J. M. (2014). L'ergonomie prospective: Fondements et enjeux. *Le Travail Humain, 77*(1), 1–20.

Brucks, M. S., & Levav, J. (2022). Virtual communication curbs creative idea generation. *Nature, 605*(7908), 108–112.

Buisine, S., & Guegan, J. (2020). Proteus vs. social identity effects on virtual brainstorming. *Behaviour & Information Technology, 39*(5), 594–606.

Burkhardt, J.-M., Détienne, F., Hebert, A.-M., Perron, L., Safin, S., & Leclercq, P. (2009). An approach to assess the quality of collaboration in technology-mediated design situations. In *ECCE '09: Proceedings of the 16th European conference on cognitive ergonomics*. ACM.

Camacho, L. M., & Paulus, P. B. (1995). The role of social anxiousness in group brainstorming. *Journal of Personality and Social Psychology, 68*, 1071–1080.

Carver, C. S., & Scheier, M. F. (1990). Origins and functions of positive and negative affect: A control-process view. *Psychological Review, 97*(1), 19.

Castelli, F. R., & Sarvary, M. A. (2021). Why students do not turn on their video cameras during online classes and an equitable and inclusive plan to encourage them to do so. *Ecology and Evolution, 11*(8), 3565–3576.

Chen, H., Liang, Q., & Zhang, Y. (2019). Facilitating or inhibiting? *Chinese Management Studies, 13*(4), 802–819.

Choi, H.-S., & Thompson, L. (2005). Old wine in a new bottle: Impact of membership change on group creativity. *Organizational Behavior and Human Decision Processes, 98*, 121–132.

Cialdini, R. B., & Trost, M. R. (1998). Social influence: Social norms, conformity and compliance. In D. T. Gilbert, S. T. Fiske, & G. Lindzey (Eds.), *The handbook of social psychology* (4th ed., pp. 151–192). McGraw-Hill.

Clerke, A. S., & Heerey, E. A. (2021). The influence of similarity and mimicry on decisions to trust. *Collabra: Psychology, 7*(1), 23441.

Coburn, J. Q., Freeman, I., & Salmon, J. L. (2017). A review of the capabilities of current low-cost virtual reality technology and its potential to enhance the design process. *Journal of Computing and Information Science in Engineering, 17*(3), 031013.

Conti, R., Amabile, T. M., & Pollak, S. (1995). The positive impact of creative activity: Effects of creative task engagement and motivational focus on college students' learning. *Personality and Social Psychology Bulletin, 21*(10), 1107–1116.

Corazza, G. E. (2016). Potential originality and effectiveness: The dynamic definition of creativity. *Creativity Research Journal, 28*(3), 258–267.

Corazza, G. E. (2023). Définition de la créativité. In N. Bonnardel, F. Girandola, E. Bonetto, & T. Lubart (Eds.), *La Créativité en Situations : Théories et Applications* (pp. 30–39). Dunod.

Cottrell, N. B. (1972). Social facilitation. *Experimental Social Psychology, 185*, 236.

Csikszentmihalyi, M. (2014). Society, culture, and person: A systems view of creativity. In M. Csikszentmihaly. *The Systems Model of Creativity*. Dordrecht: Springer.

Darbellay (Ed.). (2024). *Elgar encyclopedia of interdisciplinarity and transdisciplinarity*. Edward Elgar Publishing.

De Bono, E. (1970). *Lateral thinking* (pp. 1–32). Penguin.

De Bono, E. (1985). *Six thinking hats: An essential approach to business management from the creator of lateral thinking*. Little, Brown and Co.

De Bono, E. (1991). *Six action shoes*. Harper Collins Publishers.

De Dreu, C. K. W., Nijstad, B. A., & Baas, M. (2024). Human creativity: Functions, mechanisms, and social conditioning. *Advances in Experimental Social Psychology, 69*, 203–262.

De Dreu, C. K. W., Nijstad, B. A., & van Knippenberg, D. (2008). Motivated information processing in group judgment and decision making. *Personality and Social Psychology Review, 12*, 222–249.

De Dreu, C. K. W., & West, M. A. (2001). Minority dissent and team innovation: The importance of participation in decision making. *Journal of Applied Psychology, 86*, 1191–1201.

De Rooij, A., Van Der Land, S., & Van Erp, S. (2017, June). The creative Proteus Effect: How self-similarity, embodiment, and priming of creative stereotypes with avatars influences creative ideation. In *Proceedings of the 2017 SIGCHI conference on creativity and cognition* (pp. 232–236). ACM.

de Vasconcelos Filho, J. E., Inkpen, K. M., & Czerwinski, M. (2009, May). Image, appearance and vanity in the use of media spaces and video conference systems. In *Proceedings of the 2009 international conference on supporting group work* (pp. 253–262). ACM.

DeBruine, L. M. (2002). Facial resemblance enhances trust. *Proceedings of the Royal Society of London. Series B: Biological Sciences, 269*(1498), 1307–1312.

Delignières, D. (2016). La mesure en psychologie. *Blog Personnel*. https://dididerdelignieresblog.files.wordpress.com/2016/03/deamesure.pdf

Denstadli, J. M., Julsrud, T. E., & Hjorthol, R. J. (2012). Videoconferencing as a mode of communication: A comparative study of the use of videoconferencing and face-to-face meetings. *Journal of Business and Technical Communication, 26*(1), 65–91.

Dewett, T. (2007). Linking intrinsic motivation, risk taking, and employee creativity in an R&D environment. *R&D Management, 37*, 137–149.

Diehl, M., & Stroebe, W. (1987). Productivity loss in brainstorming groups: Toward the solution of a riddle. *Journal of Personality and Social Psychology, 53*, 497–509.

Dreisbach, G., & Goschke, T. (2004). How positive affect modulates cognitive control: Reduced perseveration at the cost of increased distractibility. *Journal of Experimental Psychology: Learning, Memory, and Cognition, 30*(2), 343.

Du, Y., Zhang, L., & Chen, Y. (2016). From creative process engagement to performance: Bidirectional support. *Leadership and Organization Development Journal, 37*(7), 966–982.

Dubois, N. (2003). *A sociocognitive approach to social norms.* Routledge.

Eastman, C. M. (1969). Cognitive processes and ill-defined problems: A case study from design. In *Proceedings of the 1st international joint conference on artificial intelligence* (pp. 669–690).

Fauville, G., Luo, M., Queiroz, A. C. M., Lee, A., Bailenson, J. N., & Hancock, J. (2023). Video-conferencing usage dynamics and nonverbal mechanisms exacerbate Zoom Fatigue, particularly for women. *Computers in Human Behavior Reports, 10*, 100271.

Fischer, I. (2009). Friend or foe: Subjective expected relative similarity as a determinant of cooperation. *Journal of Experimental Psychology: General, 138*(3), 341.

Gibson, C. B., & Gibbs, J. L. (2006). Unpacking the concept of virtuality: The effects of geographic dispersion, electronic dependence, dynamic structure, and national diversity on team innovation. *Administrative Science Quarterly, 51*(3), 451–495.

Gilovich, T., Medvec, V. H., & Savitsky, K. (2000). The spotlight effect in social judgment: An egocentric bias in estimates of the salience of one's own actions and appearance. *Journal of Personality and Social Psychology, 78*(2), 211.

Glaeser, E. L., Laibson, D. I., Scheinkman, J. A., & Soutter, C. L. (2000). Measuring trust. *The Quarterly Journal of Economics, 115*(3), 811–846.

Glaveanu, V. P. (2013). Rewriting the language of creativity: The Five A's framework. *Review of General Psychology, 17*, 69–81.

Glăveanu, V., Lubart, T., Bonnardel, N., Botella, M., de Biaisi, P.-M., Desainte-Catherine, M., Georgsdottir, A., Guillou, K., Kurtag, G., Mouchiroud, C., Storme, M., Wojtczuk, A., & Zenasni, F. (2013). Creativity as action: Findings from five creative domains. *Frontiers in Psychology, 4*, 1–14.

Goh, C. H., Kulathuramaiyer, N., & Zaman, T. (2017). Riding waves of change: A review of Personas research landscape based on the Three Waves of HCI. In *Proceedings of the international conference on social implications of computers in developing countries* (pp. 605–616). Springer.

Goncalo, J. A., Katz, J. H., & Ellis, L. M. (2019). PIECE together: How social norms support the process of team creativity. In P. B. Paulus & B. A. Nijstad (Eds.), *The Oxford handbook of group creativity and innovation* (pp. 217–230). Oxford University Press.

Goncalo, J. A., & Staw, B. M. (2006). Individualism-collectivism and group creativity. *Organizational Behavior and Human Decision Processes, 100*, 96–109.

Gorisse, G., Wellenreiter, S., Fleury, S., Lécuyer, A., Richir, S., & Christmann, O. (2023). I am a genius! Influence of virtually embodying leonardo da vinci on creative performance. *IEEE Transactions on Visualization and Computer Graphics, 29*(11), 4328–4338.

Greenaway, K. H., Wright, R. G., Willingham, J., Reynolds, K. J., & Haslam, S. A. (2014). Shared identity is key to effective communication. *Personality and Social Psychology Bulletin, 41*(2), 171–182.

Guegan, J., Buisine, S., Mantelet, F., Maranzana, N., & Segonds, F. (2016). Avatar-mediated creativity: When embodying inventors makes engineers more creative. *Computers in Human Behavior, 61*, 165–175.

Guegan, J., Lubart, T., & Collange, J. (2019). (Social) identity and creativity in virtual settings: Review of processes and research agenda. In *The Palgrave handbook of social creativity research* (pp. 191–207). Palgrave Macmillan.

Guegan, J., Segonds, F., Barré, J., Maranzana, N., Mantelet, F., & Buisine, S. (2017). Social identity cues to improve creativity and identification in face-to-face and virtual groups. *Computers in Human Behavior, 77*, 140–147.

Harvey, S., & Berry, J. W. (2023). Toward a meta-theory of creativity forms: How novelty and usefulness shape creativity. *Academy of Management Review, 48*(3), 504–529.

He, Y., Guo, Z., Wang, X., Sun, K., Lin, X., Wang, X., et al. (2022). Effects of audiovisual interactions on working memory task performance—Interference or facilitation. *Brain Sciences, 12*(7), 886.

Hoc, J. M. (2001). Towards a cognitive approach to human-machine cooperation in dynamic situations. *International Journal of Human-Computer Studies, 54*(4), 509–540.

Hoc, J. M., & Carlier, X. (2002). Role of a common frame of reference in cognitive cooperation: Sharing tasks between agents in air traffic control. *Cognition, Technology & Work, 4*, 37–47.

Inkpen, K. M., & Sedlins, M. (2011). Me and my avatar: Exploring users' comfort with avatars for workplace communication. In *Proceedings of the 2011' conference on computer supported cooperative work* (pp. 383–386). ACM.

Isaksen, S. G., Murdock, M. C., Firestein, R. L., & Treffinger, D. J. (1993). *Understanding and recognizing creativity: The emergence of a discipline.* Ablex Publishing.

Jansson, D. G., & Smith, S. M. (1991). Design fixation. *Design Studies, 12*, 3–11.

Joynes, C., Rossignoli, S., & Fenyiwa Amonoo-Kuofi, E. (2019). *21st century skills: Evidence of issues in definition, demand and delivery for development contexts (K4D emerging issues report).* Institute of Development Studies.

Karau, S. J., & Hart, J. W. (1998). Group cohesiveness and social loafing: Effects of a social interaction manipulation on individual motivation within groups. *Group Dynamics: Theory, Research, and Practice, 2*(3), 185.

Karau, S. J., & Williams, K. D. (1993). Social loafing: A meta-analytic review. *Journal of Personality and Social Psychology, 65*(4), 681–706.

Karau, S. J., & Williams, K. D. (1997). The effects of group cohesiveness on social loafing and social compensation. *Group Dynamics: Theory, Research, and Practice, 1*(2), 156.

Kim, K., & Jo, I.-H. (2020). The effect of visual Cues on cognitive load depending on self-regulation in video-based learning. In *Proceedings of the 13th international conference on educational data mining (EDM 2020)* (pp. 776–780).

Kristensson, P., & Norlander, T. (2003). The creative product and process in computer-mediated groups. *The Journal of Creative Behavior, 37*(4), 223–243.

Latané, B., Williams, K., & Harkins, S. (1979). Many hands make light the work: The causes and consequences of social loafing. *Journal of Personality and Social Psychology, 37*(6), 822.

Lellouche-Gounon, A., Gounon, A., & Bonnardel, N. (2023). Créativité des personnes âgées et de leurs aidants dans la cadre d'ateliers de créativité collaborative et participative. In N. Bonnardel, F. Girandola, E. Bonetto, & T. Lubart (Eds.), *La Créativité en Situations : Des Théories aux Applications* (pp. 313–323). Dunod.

Lengel, R. H., & Daft, R. L. (1988). The selection of communication media as an executive skill. *Academy of Management Perspectives, 2*(3), 225–232.

Levine, J. M., Choi, H.-S., & Moreland, R. L. (2019). Newcomer influence and creativity in work groups. In P. B. Paulus & B. A. Nijstad (Eds.), *The Oxford handbook of group creativity and innovation* (pp. 51–74). Oxford University Press.

Lopez-Bagousse, C., Arciszewski, T., & Bonnardel, N. (2025). Design-fiction approach as a support for creative design activities in prospective ergonomics: *An exploratory study. Ergonomics*, 1–17. https://doi.org/10.1080/00140139.2025.2512425

Lopez-Bagousse, C., Thomas, A., & Bonnardel, N. (2024). Supporting human factors and ergonomics professionals in anticipating future work changes in response to the anthropocene. *Ergonomics, 68*(6), 832–843. https://doi.org/10.1080/00140139.2024.2383947

Lü, L., Li, F., Leung, K., Savani, K., & Morris, M. W. (2017). When can culturally diverse teams be more creative? The role of leaders' benevolent paternalism. *Journal of Organizational Behavior, 39*(4), 402–415.

Lubart, T. I. (1994). Creativity. In R. J. Sternberg (Ed.), *Thinking and problem solving* (pp. 289–332). Academic Press.

Lubart, T. I. (2017). The 7 C's of creativity. *The Journal of Creative Behavior, 51*, 293–296.

Lubart, T. I., Mouchiroud, C., Tordjman, S., & Zenasni, F. (2015). *Psychologie de la Créativité* (2nd ed.). Armand Colin.

Maciver, F., & Malins, J. (2016). Two heads are better than one: Principles for collaborative design practice. In *Collaboration in creative design: Methods and tools* (pp. 13–31). Springer International Publishing.

Marinussen, M., & de Rooij, A. (2019). Being yourself to be creative: How self-similar avatars can support the generation of original ideas in virtual environments. In *Proceedings of the 2019 on creativity and cognition C&C '19* (pp. 285–293). ACM.

McMahon, K., Ruggeri, A., Kämmer, J. E., & Katsikopoulos, K. V. (2016). Beyond idea generation: The power of groups in developing ideas. *Creativity Research Journal, 28*(3), 247–257.

Mercier, M., Vinchon, F., Pichot, N., Bonetto, E., Bonnardel, N., Girandola, F., & Lubart, T. (2021). COVID-19: A boon or a bane for creativity? *Frontiers in Psychology, 11*, 3916.

Milliken, F. J., Bartel, C. A., & Kurtzberg, T. R. (2003). Diversity and creativity in work groups. In *Group creativity* (pp. 32–62). Oxford University Press.

Murrar, S., & Brauer, M. (2024). Using social norms to promote positive relations between social groups. In E. Halperin, B. Hameirii, & R. Littman (Eds.), *Psychological intergroup interventions: Evidence-based approaches to improve intergroup relations* (pp. 45–68). Routledge.

Nemeth, C. J., & Nemeth-Brown, B. (2003). Better than individuals? The potential benefits of dissent and diversity for group creativity. In *Group creativity: Innovation through collaboration* (pp. 63–84). Oxford University Press.

Nemeth, C. J., & O'Connor, K. (2019). Better than individuals? Dissent and group creativity. In P. B. Paulus & B. A. Nijstad (Eds.), *The Oxford handbook of group creativity and innovation* (pp. 75–92). Oxford University Press.

Nemeth, C. J., & Ormiston, M. (2007). Creative idea generation: Harmony vs. stimulation. *European Journal of Social Psychology, 37*(3), 524–535.

Nemeth, C. J., Personnaz, B., Personnaz, M., & Goncalo, J. A. (2004). The liberating role of conflict in group creativity: A cross-cultural study. *European Journal of Social Psychology, 34*, 365–374.

Nijstad, B. A., Bechtold, M., & Choi, H.-C. (2019). Information processing, motivation, and group creativity. In P. B. Paulus & B. A. Nijstad (Eds.), *The Oxford handbook of group creativity and innovation* (pp. 87–102). Oxford University Press.

Nijstad, B. A., Diehl, M., & Stroebe, W. (2003b). Cognitive stimulation and interference in idea-generating groups. In P. B. Paulus & B. A. Nijstad (Eds.), *Group creativity: Innovation through collaboration* (pp. 137–159). Oxford University Press.

Nijstad, B. A., & Stroebe, W. (2006). Cognitive model of idea generation in groups. *Personality and Social Psychology Review, 10*(3), 186–213.

Nijstad, B. A., Stroebe, W., & Lodewijkx, H. F. (2003a). Production blocking and idea generation: Does blocking interfere with cognitive processes? *Journal of Experimental Social Psychology, 39*(6), 531–548.

Nowak, K. L., & Fox, J. (2018). Avatars and computer-mediated communication: A review of the definitions, uses, and effects of digital representations. *Review of Communication Research, 6*, 30–53.

OECD. (2019). *Stratégie 2019 de l'OCDE sur les compétences: Des compétences pour construire un avenir meilleur*. OECD.

Osborn, A. F. (1953/1963). *Applied imagination*. New York' Scribner's.

Paluck, E. L., & Shah, A. K. (2025). Field research: Five approaches to theoretical development in the field. In D. T. Gilbert, S. T. Fiske, E. J. Finkel, & W. B. Mendes (Eds.), *The handbook of social psychology* (6th ed.). Situational Press.

Park, H., Brown, S., & Chu, S. L. (2019). Understanding avatar identification through visual similarity for richer story creation. In *HCI international 2019-posters: 21st international conference, HCII 2019, Orlando, FL, USA, July 26–31, 2019, Proceedings, Part II 21* (pp. 423–431). Springer International Publishing.

Pellegrin, L., Chassery, L., Chaudet, H., Texier, G., & Bonnardel, N. (2021). Decision-making during nonroutine outbreak management: Toward an exploration of experts' creative decisions. *Applied Ergonomics, 90*, 103232.

Postmes, T., Spears, R., & Cihangit, S. (2001). Quality of decision making and group norms. *Journal of Personality and Social Psychology, 80*, 918–930.

Pruitt, J., & Grudin, J. (2003). Personas: Practice and theory. In *Proceedings of the conference on designing for user experiences* (pp. 1–15). ACM.

Purcell, A. T., & Gero, J. S. (1996). Design and other types of fixation. *Design Studies, 17*(4), 363–383.

Ratan, R., Beyea, D., Li, B. J., & Graciano, L. (2020). Avatar characteristics induce users' behavioral conformity with small-to-medium effect sizes: A meta-analysis of the proteus effect. *Media Psychology, 23*(5), 651–675.

Ratan, R., Klein, M. S., Ucha, C. R., & Cherchiglia, L. L. (2022). Avatar customization orientation and undergraduate-course outcomes: Actual-self avatars are better than ideal-self and future-self avatars. *Computers & Education, 191*, 104643.

Rhodes, M. (1961). An analysis of creativity. *Phi Delta Kappan, 42*, 305–310.

Rietzschel, E. F., Nijstad, B. A., & Stroebe, W. (2006). Productivity is not enough: A comparison of interactive and nominal brainstorming groups on idea generation and selection. *Journal of Experimental Social Psychology, 42*(2), 244–251.

Rietzschel, E. F., Nijstad, B. A., & Stroebe, W. (2010). The selection of creative ideas after individual idea generation: Choosing between creativity and impact. *British Journal of Psychology, 101*(*1*), 47–68.

Robert, J. M., & Brangier, E. (2012). Prospective ergonomics: Origin, goal, and prospects. *Work, A Journal of Prevention, Assessment & Rehabilitation, 41*, 5235–5242.

Runco, M. A., & Jaeger, G. J. (2012). The standard definition of creativity. *Creativity Research Journal, 24*(*1*), 92–96.

Safin, S., Détienne, F., Burkhardt, J. M., Hebert, A. M., & Leclercq, P. (2021). The interplay between quality of collaboration, design project evolution and outcome in an architectural design studio. *CoDesign, 17*(4), 392–409.

Sawyer, R. K. (2007). *Group genius: The creative power of collaboration.*

Schneider, S., Beege, M., Nebel, S., & Rey, G. D. (2018). A meta-analysis of how signaling affects learning with media. *Educational Research Review, 23*, 1–24.

Scott, C. L. (2015). *[The futures of learning : Why must learning content and methods change in the 21st century?, par UNESCO]*. Les Apprentissages de Demain 2 : Quel Type d'apprentissage Pour Le XXIe Siècle ?, 14, 18.

Shih, M. T., Lee, Y. C., Huang, C. M., & Chan, L. (2023). A feeling of "déjà vu": The effects of avatar appearance-similarity on persuasiveness in social virtual reality. *Proceedings of the ACM on Human-Computer Interaction, 7*(CSCW2), 1–31. ACM.

Shiue, Y. C., Chiu, C. M., & Chang, C. C. (2010). Exploring and mitigating social loafing in online communities. *Computers in Human Behavior, 26*(4), 768–777.

Simon, H. A. (1973). The structure of ill structured problems. *Artificial Intelligence, 4*(3-4), 181–201.

Spoelma, T., & Ellis, A. P. J. (2017). Fuse or fracture? Threat as a moderator of the effects of diversity faultlines in teams. *Journal of Applied Psychology, 102*(9), 1344–1359.

Stasser, G., & Abele, S. (2020). Collective choice, collaboration, and communication. *Annual Review of Psychology, 71*(1), 589–612.

Steiner, I. D. (1972). *Group process and productivity*. Academic Press.

Sternberg, R. J., & Lubart, T. I. (1999). The concept of creativity: Prospects and paradigms. In R. J. Sternberg (Ed.), *Handbook of creativity* (pp. 3–15). Cambridge University Press.

Suwa, M., Gero, J. S., & Purcell, T. A. (1998a, May). The roles of sketches in early conceptual design processes. In *Proceedings of the twentieth annual conference of the cognitive science society* (Vol. 20, pp. 1043–1048). Routledge.

Suwa, M., Purcell, T., & Gero, J. (1998b). Macroscopic analysis of design processes based on a scheme for coding designers' cognitive actions. *Design Studies, 19*(4), 455–483.

Tajfel, H., & Turner, J. C. (1979). An integrative theory of intergroup conflict. In W. G. Austin & S. Worchel (Eds.), *The social psychology of intergroup relations* (pp. 33–37). Brooks/Cole.

Thornhill-Miller, B., Camarda, A., Mercier, M., Burkhardt, J.-M., Morisseau, T., Bourgeois-Bougrine, S., Vinchon, F., El Hayek, S., Augereau-Landais, M., Mourey, F., Feybesse, C., Sundquist, D., & Lubart, T. (2023). Creativity, critical thinking, communication, and collaboration: Assessment, certification, and promotion of 21st century skills for the future of work and education. *Journal of Intelligence, 11*(3), 54.

Toumi, K., Girandola, F., & Bonnardel, N. (2021). Technologies for supporting creativity in design: A view of physical and virtual environments with regard to cognitive and social processes. *Creativity: Theories – Research – Applications, 8*(1), 189–212.

Treffinger, D. J. (1995). Creative problem solving: Overview and educational implications. *Educational Psychology Review, 7*(3), 301–311.

Tromp, C. (2022). Creativity from constraint exploration and exploitation. *Psychological Reports.*

Turner, J. C., & Oakes, P. J. (1986). The significance of the social identity concept for social psychology with reference to individualism, interactionism and social influence. *British Journal of Social Psychology, 25*(3), 237–252.

Van Der Land, S. F., Schouten, A. P., Feldberg, F., Huysman, M., & van den Hooff, B. (2015). Does avatar appearance matter? How team visual similarity and member–avatar similarity influence virtual team performance. *Human Communication Research, 41*(1), 128–153.

van Knippenberg, D., De Dreu, C. K. W., & Homan, A. C. (2004). Work group diversity and group performance: An integrative model and research agenda. *Journal of Applied Psychology, 89*(6), 1008–1022.

van Knippenberg, D., Li, J., & Tu, Y. (2024). Team informational resources, information elaboration, and team innovation: Diversity mindset moderating functional diversity and boundary spanning scouting effects. *Journal of Occupational and Organizational Psychology, 97*(4), 1835–1853.

Vincent-Lancrin, S., Gonzalez-Sancho, C., Bouckaert, M., de Luca, F., Fernández-Barrerra, M., Jacotin, G., Urgel, J., & Vidal, Q. (2019). *Fostering students' creativity and critical thinking: What it means in school.* OECD Publishing.

Ward, T. B., Patterson, M. J., Sifonis, C. M., Dodds, R. A., & Saunders, K. N. (2002). The role of graded category structure in imaginative thought. *Memory & Cognition, 30*, 199–216.

Witt, K. J., Oliver, M., & McNichols, C. (2016). Counseling via avatar: Professional practice in virtual worlds. *International Journal for the Advancement of Counselling, 38*, 218–236.

Wu, L., Wang, D., & Evans, J. A. (2019). Large teams develop and small teams disrupt science and technology. *Nature, 566*, 378–382.

Xie, H., Wang, F., Hao, Y., Chen, J., An, J., Wang, Y., & Liu, H. (2017). The more total cognitive load is reduced by cues, the better retention and transfer of multimedia learning: A meta-analysis and two meta-regression analyses. *PLoS One, 12*(8), e0183884.

Yang, Y., & Konrad, A. M. (2010). Diversity and organizational innovation: The role of employee involvement. *Journal of Organizational Behavior, 32*(8), 1062–1083.

Yee, N., & Bailenson, J. (2007). The Proteus effect: The effect of transformed self-representation on behavior. *Human Communication Research, 33*(3), 271–290.

Yung, H. I., & Paas, F. (2015). Effects of cueing by a pedagogical agent in an instructional animation: A cognitive load approach. *Journal of Educational Technology & Society, 18*(3), 153–160.

Zhu, R., & Yi, C. (2024). Avatar design in Metaverse: The effect of avatar-user similarity in procedural and creative tasks. *Internet Research, 34*(1), 39–57.

Ziegler, C. N., & Golbeck, J. (2007). Investigating interactions of trust and interest similarity. *Decision Support Systems, 43*(2), 460–475.

CHAPTER 2

Circumscribing and Appraising Collaboration in Design

Michael J. Baker, Jean-Marie Burkhardt, Françoise Détienne, and Stéphane Safin

Introduction

Amongst the range of collaborative activities, design plays a particular role. It is a goal-oriented activity that includes an intrinsic creative dimension, aiming at the production of artefacts, whether unique or

M. J. Baker (✉) • F. Détienne
Télécom Paris, Institut Interdisciplinaire de l'Innovation (i3), CNRS UMR 9217, Paris, France
e-mail: michael.baker@telecom-paris.fr; francoise.detienne@telecom-paris.fr

J.-M. Burkhardt
Université Gustave Eiffel and Université Paris Cité, LaPEA, Versailles, France
e-mail: jean-marie.burkhardt@univ-eiffel.fr

S. Safin
Télécom Paris, Institut Interdisciplinaire de l'Innovation (i3), CNRS UMR 9217, Paris, France

Université Libre de Bruxelles, Faculty of Psychological Science and Education, Research Center for Work and Consumer Psychology, Brussels, Belgium
e-mail: stephane.safin@ulb.be

N. Bonnardel (ed.), *Collaborations in Creativity*, The Seven C's of Creativity, https://doi.org/10.1007/978-3-032-07753-0_2

mass-produced, intended for one or more uses. The term "design" itself is ambiguous. It can be understood as a project process, where it consists of all the steps from a request to the delivery of an artefact. Alternatively, the term can be understood as a social and cognitive activity, primarily occurring in the early stages of a project, during which a problem is (re)defined, an original and relevant solution to this problem is imagined, and this solution is specified so that it can thereafter be subject to production (construction or machining). In this chapter, we refer to this second meaning.

Design and creativity are objects of numerous definitions in the literature, and most definitions of the two concepts emphasise the idea of a thinking process that leads to the production of concepts that are both original and contextually appropriate (e.g., Akin & Akin, 1998; Simon, 1995; Lubart, 1994; Bonnardel, 2009). However, one can argue, with Richard (2016), that there is a form of continuum between these two notions, with creativity focusing more on originality, and design on the adaptation to context. Moreover, the early stages of design, during which concepts are developed, are often referred to as the creative stages of design.

In this chapter, as a follow-up to Chap. 1 (by Bonnardel et al.), we discuss the key characteristics of collaborative design, as it has been studied by numerous scholars in different disciplines. Far from conceptualising design as a 'mechanical' and predictable activity, we adopt a situated and constructivist viewpoint: the design activity is developed by actors grounded in singular social and cultural situations. We position collaboration in relation to other forms of collective organisation, discussing the multidimensional nature of collaboration in design, particularly how it articulates discursive, instrumental, affective, and organisational dimensions, and how it is anchored in systems of values and cultures. We identify the challenges of characterising and evaluating the quality of collaboration in design in real situations and conclude this chapter with perspectives for research.

Collective Activity, Coordination, Cooperation and Collaboration

The concept of collaboration, designating a specific form of activity involving human, non-human and perhaps also artificial agents,[1] has become the focus of several (inter-)disciplinary areas of research over the last fifty years—for example, in computer-supported cooperative work (CSCW), conversation analysis, human–computer dialogue, computer supported collaborative learning (CSCL), collaborative creativity and collaborative design. This no doubt relates to societal and economic changes, such as the erosion of authority in the home, the school and the workplace, the emergence of new forms of participative democracy, changes in work practices beyond Taylorism, globalisation that requires cross-cultural team work, and, concomitantly, the rise of Internet technologies and social networks.

Unsurprisingly, there is little agreement on what collaboration is across these areas of research. We therefore aim to simply present an internally coherent definition that is fruitful for research (cf. Andriessen & Baker, 2020). In the first instance, collaboration needs to be defined with respect to a set of related concepts, including collective activity, coordination and cooperation.

A collective activity is something that corresponds to a form of organisation oriented towards satisfying a societal need, such as having shelter, food and protection from others and the elements. Leont'ev (1979) discussed the famous example of hunting. This requires division of labour, between people making weapons (and those obtaining the raw materials for it), people preparing hunting grounds, beating the prey in a given direction, killing the prey, skinning and cutting it up, and so on. These activities need to be coordinated in time (you need the weapons before the hunt). A more recent example would be Wikipedia: all involved are working to facilitate knowledge acquisition on a global scale, without there necessarily being actual synchronous interaction between participants. Their orientation is not always or necessarily towards other participants, but rather towards the object of the activity (knowledge production).

[1] See the special issue of the journal *Interaction Studies: Social Behaviour and Communication in Biological and Artificial Systems*, 16(3), 2015, on "Coordination, Collaboration and Cooperation: Interdisciplinary Perspectives": https://www.jbe-platform.com/content/journals/15720381/16/3

Coordination is, according to the literal meaning of the word, the organisation of the *ordering* of activities, actions, behaviours in time. This can occur on different temporal levels, from the order in which activities are carried out (see above) to managing utterances in dialogue, turn-taking. Coordination on the level of actions, in order to be necessary, presupposes cooperative activity.

Cooperation, as a concept, is defined primarily in terms of there being one or more *goals* that are believed to be shared, towards which all involved parties orient their actions. As with collective activity, people who cooperate do not necessarily need to orient towards others in real-time, since it is possible to divide up sub-tasks between people. It is, however, necessary to *coordinate* individual or sub-group activities on a more local, small group scale, so that they move towards the shared goal. On another level, cooperating implies not blocking others' actions provided that they do not impede joint progress, and even helping others, going beyond what they ask for explicitly.

Finally, in this conceptual nexus, collaboration emerges as a specific way in which small groups may work together, cooperate, on specific tasks, in real time. This presupposes a general collective activity framework (e.g. a private or public organisation's aims), coordination on the level of a localised interpersonal interaction, and a shared cooperative goal. *Collaboration is a particular way of conducting cooperation*, where participants strive not only to achieve a shared task, but also shared understanding of what they are doing together. Cooperation, however, presupposes at least some degree of collaboration, for example, in order to define the shared goal and to assemble results of individual or sub-group work into a coherent whole. Consider the example of four people 'working together' (we deliberately use a general term here) to design the decoration of a communal leisure space in their office. They might collaborate in defining an overall set of objectives (e.g. comfortable, inexpensive to decorate, easy to clean, a place for shared meals) then cooperate by dividing up responsibilities (who decides the furniture, the decoration, the artefacts necessary, such as a microwave, a fridge, …). Once each has found solutions to their assigned sub-tasks, it would be necessary to collaborate, discussing the ensemble to check coherence.

The nexus of concepts surrounding collaboration is represented in Fig. 2.1.

Figure 2.1 represents collective activity as the broadest category relating to collaboration, defined on a societal level in terms of a basic motive

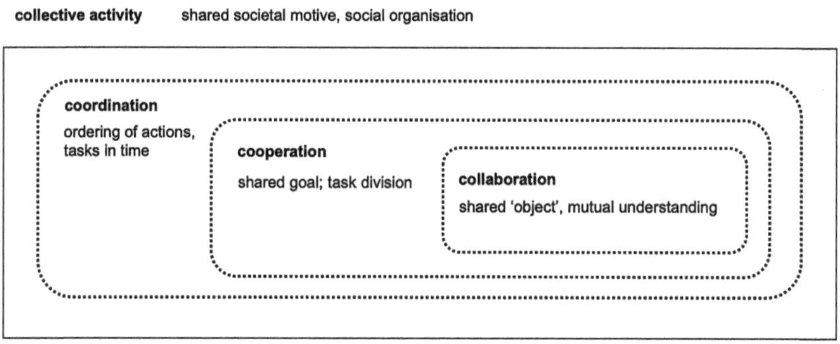

Fig. 2.1 Collaboration defined in relation to collective activity, coordination and cooperation

(such as providing shelter, or designing and producing tools that improve efficiency of work, improve comfort, facilitate transport, etc.), including division of labour and coordination between related activities. On the next level, each sub-activity (such as making tools, applying them, distributing products, etc.) requires coordination both with other related sub-activities and within itself, on the level of ordering of actions, including utterances, in time. A coordinated activity requires cooperation, in the minimal sense of jointly pursuing a shared goal, that may be carried out by dividing up responsibilities for sub-tasks amongst participants. Moreover, a coordinated cooperative activity may be achieved by collaboration, that requires a joint focus and common agency towards elaborating a shared representation of the task.

Finally, the concept of collaboration is intimately associated with values. Collaboration itself, as a process, can be seen as embodying the positive value of mutual respect for the ideas and feelings of others. And yet, it is possible to collaborate with respect to an objective that could be seen as involving a negative value judgement (e.g. sharing the ideology of an invading enemy, planning how to dominate or harm others).

We shall deal with these issues in more detail below.

Collaboration Implies a Shared Task Focus

Whereas both cooperation and collaboration imply at least a goal that is (mutually assumed to be) shared, with a task distribution setting, collaboration goes further than this, in implying a *shared task focus*. In collaboration, task and problem definition are seen as co-constructed and negotiated: *grounding* (Clark & Schaefer, 1989) is indeed necessary to ensure mutual understanding of shared focus. In this direction, a collaboration can be seen as *a continued and synchronous attempt to construct and maintain a shared representation of the problem* to be solved (Roschelle & Teasley, 1995; Baker, 2015). In collaborative design, in particular in the case of multidisciplinary teams, this relates to the notion of *interdesigners' compatible representations* (Visser, 2006): compatibility in this sense does not presuppose a kind of match between what is 'shared'. Furthermore, it should be recalled that the instability of design requirements and the co-evolution of problem-solutions are inherent to design activities. In this context of open and ill-defined design tasks, co-elaborating and maintaining a shared task focus are of primordial importance to collaboration.

Having a shared task focus relates to *positive interdependency* (Johnson & Johnson, 1994) between solution elements and contributions to a shared overall task, whether divided up or not. This converges in our sense with the notion of mutual dependence in work conceived as a positive interdependency, although not always harmonious (Schmidt, 1994). Mutual interdependency has been proposed to be further decomposed into task interdependence and output interdependence, whereas task interdependence refers to "the interconnections among tasks such that the performance of one definite piece of work depends on the completion of other definite pieces of work, and output interdependence and outcome interdependence is defined as the extent to which team members' personal benefits and costs depend on successful goal attainment by other team members" (Van den Bossche et al., 2006).

Collaboration Implies a Certain Degree of Symmetry

The notion of collaboration presupposes a certain degree of equality between participants in terms of rights to contribute (notwithstanding a more or less hierarchical situation), in the context of a socio-institutional mode of organisation that favours co-elaboration of ideas, knowledge objects or tangible artefacts. For example, we would argue that it would not be useful to assimilate under the concept of collaboration situations where there is a 'master and apprentice', or when one participant has the institutionalised role of completely directing the work of others. Symmetry does not mean that everyone is the same (which is obviously false): as we said, it is about equality of *rights* to intervene in an equal manner in the ongoing activity. This means that certain social situations are not, on the face of it, places of collaboration. For example, in classical education, the teacher has the right to negatively evaluate the students' responses, but the opposite is rarely the case.

Although less covered in previous studies on design as well as in studies on collaborative design, the degree of symmetry in contributions of participants in design interactions can be understood in terms of *cooperative orientation*. This aspect is similar to the notion of reciprocal interaction highlighted by Spada et al. (Spada et al., 2005) and *symmetry/asymmetry of roles*, spontaneously adopted and negotiated in the interaction (e.g. each participant being 'proposer' of solutions; certain adopting 'critiquer' roles, explainer/explainee,...) (Baker, 2002).

In a collaborative design situation, symmetry may be understood with respect to the dimension of *interactive roles* emerging from interactions (see Détienne et al., 2016). On a dialogic dimension, we expect some kind of symmetry in informative and regulative roles spontaneously adopted by participants (based on regular exchanges, e.g., between who provides idea and who criticises and extends ideas) together with asymmetry in regulative roles (group management as observed, e.g., in interactions in OSS design projects: Barcellini et al., 2014). In multidisciplinary design, collaboration may be based on some degree of symmetry on the epistemic dimension, giving a basis for a (partial) common frame of reference, with asymmetry between respective epistemic domains of reference highlighting complementarity in specialised knowledge of contributors (as observed, e.g., by Baker et al., 2009; Loup-Escande et al., 2014).

Collaboration Requires Regulation, Coordination and Organisation

Collaboration requires coordination and organisation within a socio-technical system, involving procedures, methods, tools, constraints (time, cost); that is the design collaborative task, in its prescribed or effective forms. This latter distinction, classical in cognitive ergonomics, refers, respectively, to a descriptive view of the goals and resources, as well as to the way the actors themselves represent them (Leplat, 2004). These methods and tools will ultimately affect the way the group's members act in order to carry out the task, as well as the collaboration itself.

In fact, the achievement of a design task, and collaboration with respect to it, may each be more or less constrained by rules and procedures, depending on the situation. These can be termed, respectively, 'task-scripts' and collaboration scripts' (e.g. Engeström et al., 1997). For example, certain tasks may be governed by more or less well-established and explicit procedures that have to be followed (task script), such as ordering topics of discussion (general layout before interior design, before choice of furniture). Similarly, collaboration can occur between the extremes of completely 'free' discussion and situations where specific stages of the collaboration are specified (e.g. free uncritical idea generation, followed by selection, organisation, synthesis and decision-making), and even 'ground rules' for communication (e.g. 'try to listen to and respond constructively to each other's ideas') within these stages.

Methods and tools can also structure the collaboration itself. Taking up the brainstorming example, a collaboration script could encourage each designer to produce at least *n* ideas, or to shift regularly from a role of idea proposer to a role of idea critic, regulating symmetry of roles. When applied to situations with a learning objective, a central issue is also how to design communication interfaces and shared learning task spaces that 'structure' or 'script' collaboration in ways that are designed to be optimal for learning (see e.g. Fischer et al., 2007). Collaboration can be structured in several ways, for example by constraining the set of available speech acts (Baker & Lund, 1997), by imposing a strict task sequence, by providing specific semiotic means of expression, such as argumentation dialogue diagrams (see Andriessen et al., 2003). The possibility of 'over-scripting' has been debated (Dillenbourg, 2002), or rather the extent to which collaboration can and should be scripted, so as to favour optimal forms of collaboration yet without stifling the essential creativity of dialogue. These issues are also discussed in Chap. 1 (Bonnardel et al.) and Chap. 3 (Bonnardel & Girandola) of this book.

Collaboration Is Mediated

Every collaborative design project involves the use of mediating instruments for collaboration. On one hand, design situations are varied: while face-to-face meetings are common, collective design activities are also organised remotely and/or asynchronously, requiring the use of tools for communication and coordination. On the other hand, design is an activity that relies entirely on the production of representations of the object to be designed (Visser, 2006). Creating these representations always involves the use of graphic expression instruments, ranging from rudimentary tools such as paper and pencil to the most complex computer-aided modelling software, as well as tangible prototyping tools. Representation and communication tools impact, structure, support, and constrain the creative and collective processes involved in design. They mediate epistemic relationships (understanding the characteristics of the artefact), pragmatic relationships (actions of transformation on this artefact) to the object being designed, and social and collaborative processes required for collective design.

From the perspective of creative processes, the graphic representations of the designed artefact serve multiple roles (Safin, 2025): as supports for externalising the designers' thoughts, they support processes of verifying the coherence of the artefact and allow the emergence of unexpected discoveries (Suwa et al., 2000) through reflective conversation with the situation (Schön, 1983). Representation systems allow designers to choose the best viewpoints on the object to infer its characteristics (Rietveld & Brouwers, 2017) and partly structure the organisation of the design activity, providing spaces for exploring solutions and spaces for evaluation processes (Safin et al., 2011).

From a collaborative perspective, graphic expression tools not only support the collective activity of generating and evaluating solutions but also take on roles of coordination and management of interactions (Safin et al., 2012). Thus, representations serve as intermediate objects of design (Mer et al., 1995), mediating relationships between partners: they constitute both a support for thought and creation, a snapshot of the evolution of the designed object, and a memory of the ongoing design process. By clarifying certain ambiguities and supporting an explicit understanding of the object, they allow participants to build a shared vision of the object to be designed, contributing to the cognitive synchronisation of the various actors in the project (Falzon, 1994), the construction of a common ground (Clark & Brennan, 1991), and shared knowledge (Beers et al., 2006) among stakeholders, which is necessary for high quality collaboration.

Furthermore, representations act as boundary objects (Star & Griesemer, 1989), supporting the translation processes between different disciplines. By gathering information on common supports, stakeholders—whose skills may sometimes be distant from each other—are able to express themselves in a common language, to build a mutual understanding, and to articulate their complementary viewpoints on the project, which are necessary in integrated collective design (Détienne et al., 2005). Shared graphic representations also simplify communication: anything that cannot be drawn or pointed out must be explicitly stated.

Finally, representations and tools are also mobilised by designers for managing communicative interactions in collective design activities. Representations, used during synchronous meetings, support a high degree of interactivity among partners (Heiner et al., 2004; Safin et al., 2012). In conjunction with verbal content and gestural practices such as pointing, sharing representations allows for the establishment of a common reference context and facilitates a joint focus of collaborators' attention on specific elements of the project.

Collaboration Involves Discursive, Socio-Relational and Affective Processes

Although designers' knowledge heuristics are essential for design, designing is 'more than a cognitive process': it is also a social process in which solutions are negotiated (Bucciarelli, 1988). Such a social process of negotiation in collaborative design goes beyond cognitive design processes in terms of three other main types of processes: discursive, socio-relational and affective.

Discursive processes are the means by which participants in a collaboration use language to 'do work on' representations, for example, by elaborating each other's proposals/ideas, drawing inferences from them, giving them argumentative or explanatory foundations. This is more than the pooling of individual designers' knowledge in that collaborative design solutions *emerge* from designers' interactions in a way that goes beyond the sum of individual contributions. Partly inspired by work on rhetorical relations (Mann & Thompson, 1988) and on discursive operations

(Vignaux, 1990), it is possible to define a set of operations working on a linguistic-cognitive plane, that characterise knowledge co-elaboration, either within a specific participant's contributions or, more importantly, between participants' contributions (Baker, 1994). These include operations that provide foundations for proposals (justification, explanation, argumentation), those that elaborate or extend proposals, that restrict them and reformulate them.

However, such discursive operations produced in social interaction always carry with them an aspect of face preservation or threat (Brown & Levinson, 1987): any transformation of another's ideas, in collaborative dialogue, constitutes also a certain *evaluation* of those ideas and of the person who proposed them, with all the affective aspects which will necessarily be associated. This is particularly apparent in the case of verbal conflict and argumentation dialogue. Any argumentative critique or "attack" (Barth & Krabbe, 1982) on a statement proposed by a participant will also be experienced by that person as, to some extent, an attack on themself, in terms of perceptions of competence that are projected. It is well established that the degree of aggressiveness of a critique is a determining aspect of how argumentation dialogue unfolds: the more the critique is perceived as aggressive (e.g. claim of irrelevance) the more it is likely that proposers of solutions will defend their views, and thereby themselves (Muntigl & Turnbull, 1998) rather than attacking their opponents' views and persons.

More generally, it has been argued (Andriessen et al., 2011) that within particular organisational settings, continued collaboration across several meetings requires establishing and maintaining a 'collaborative working relation'. This is a specific relationship relating to the context of work, which may draw on, reaffirm or pose a threat to out-of-work relationships. A crucial aspect of this is the way in which 'tensions' arising in the collaborative encounter are regulated and 'relaxed' (Bales, 1950). Indeed, each collaborating group will have its own specific tension/relaxation regulation 'signature', some collaborative working relations being able to support large leaps between tension and relaxation, others not (e.g. people who have a strong interpersonal relation can bear to have major disagreements; if not, the interaction may proceed in a more 'flat' tension/relaxation pattern). An important question concerns how such regulation interacts with collaboration on a discursive level.

In collaborative design situations, argumentative dialogues may proceed by a subtle interplay between arguments drawing on different types of knowledge and the images of their respective competencies that design participants project in the interaction (Fréard et al., 2010). When more than one participant has the same competence (e.g. interior designer), this can lead to competition and (relational) conflict, just as much as to mutual understanding and co-construction: being on the same epistemic 'ground' leads to a particular kind of close collaboration, that is different from the case where there is a group of participants each of whom has unique specialities. Thus, the way collaboration functions may go beyond knowledge-based discussion and argumentation, to involve negotiation of competencies of stakeholders. On a social level, important aspects relate indeed to mutual respect, trust, as well as to responsibilities and accountability, within situational rules and norms. More generally, "ethical consideration" for others (Allwood et al., 2000) involving mutual trust and respect can be seen as not an 'optional extra' to collaboration, but as a very condition for its continuation.

In sum, collaboration in design, seen as the joint cognitive elaboration of ideas in language-based communication (including gestural and paralinguistic aspects), is inseparable from the interplay of personalities, persons, images of self and other, with all the affects that are involved.

The study of collaborative creativity workshops published in Baker et al. (2020) illustrates the interweaving of the discursive processes of argumentation and the joint regulation of affect (see also Baker et al., 2013). In collaboration with professors working on a specialised creative design Masters course at the London College of Art (co-organised with Imperial College), training workshops were designed on the basis of going against the prevailing irenic *doxa* whereby 'every idea is a good idea', to explore the contribution of interpersonal conflicts and tensions. This approach also recognises the contribution of studies that show that conflict can be productive in design, provided that it is collaborative rather than adversarial (Badke-Schaub & Buerschaper, 2001; Badke-Schaub et al., 2010; Yong et al., 2014). One such workshop is called 'argument clinic' (Baker et al., 2020): groups of four students work on the definition of their ongoing joint creative projects, by alternating, in pairs, *pro* and *contra* roles with respect to their own projects. Group work was video-recorded, and group reflection (auto-confrontation) sessions were organised, where participants were invited to identify and discuss 'key moments' during their argument clinic sessions. Groups mostly identified 'eureka!' moments, when the argumentative process led to shared recognition that the project objectives had been redefined in a useful manner. For example, in one group, whose project was on 'touch communication', the group arrived at the idea, at the outcome of an argumentative sequence, that 'touch' should not be understood literally, but rather as a matter of sensing co-presence. Such events were usually associated with the circulation of positive affect (see also Safin et al., 2018). Overall, the students stated that the alternating argumentative role-play approach enabled them to 'think around' their project in a beneficial way.

Collaboration Is a More or Less Constructive, Productive, Creative and Reflexive Process

Although cooperation can proceed by dividing up sub-task responsibilities then combining solutions in a joint product, collaboration itself implies co-construction of the meaning of the problem to be resolved and its solutions, implying 'along the way', the elaboration of new knowledge, (mutual) understanding and social awareness (understanding of the group itself, how it works and what it is about). The distinction between 'productive' and 'constructive' collaborative interactions (e.g. Miyake, 1986; Baker, 1999) relates closely to two possible foci for the quality of collaboration. A 'productive' interaction is one that is seen, from group-internal or external points of view, as having led to a 'good' *outcome* (e.g. learning) or *product*, irrespective of how that was achieved (e.g. it could have been the result of an interaction that was highly directed by one participant 'manager', with little collaboration). To speak of a *constructive* interaction is to focus on the *process* of collaboration itself, and its quality in terms of the extent to which mutual understanding is achieved, whether participants value others' contributions, the quality of the general affective atmosphere, the extent to which participants build on and elaborate *others'* contributions rather than their own (a juxtaposition of monologues) and more generally, the knowledge that emerges from the interaction (whether or not this is reflected in the quality of the final product). A crucial question here is thus the relations between the collaboration process (constructive interaction) and product (productive interaction). In the ideal case, a constructive interaction is also a productive one; but this is not necessarily the case (an interaction can be constructive without being productive, and vice versa). In this context, it is important to bear in mind the *temporal* dimension, since a given collaborative encounter may be constructive with respect to the group, but its productivity—once the group has learned to work together well—may only be made manifest in subsequent collaborations, occurring at a much later date.

Specific forms of socio-cognitive conflict may relate to constructive argumentation (Baker, 1999), involving reflection on the degree of acceptability of proposed solutions. In fact, engaging in argumentation with a view to cooperatively resolving a conflict of ideas can be one of the most intense and constructive forms of collaboration, given that participants are often obliged to more precisely define key concepts during a debate (Walton, 1992). The study of the degree of constructivity of a collaborative interaction is one way of approaching its creativity, seen as the co-elaboration of new ideas that are feasible and socially valued (Bonnardel, 2006; Kristensson et al., 2004).

Reflexive activity, in the sense of reflecting not on the content of a problem or solution, but on one's collaboration and problem-solving strategies, may lead teams to consciously realise that they are stuck or are not efficient and therefore to change and improve their practices. This refers to the distinction in cognitive ergonomics between functional and meta-functional activity (Falzon, 1994). Schön (1983) stated that self-reflection is the key to successful designing for both individuals and teams. The beneficial effects of self-reflection in design teams have been emphasised by many authors (e.g. Badke-Schaub 1999; Valkenburg, 2000) and have been described as powerful in dealing with complexity. However, Badke-Schaub (1999) also noted that self-reflection remains rare and, when criticisms of the team's approach occurred, the reactions of the team were mostly unfavourable.

Culture and Value

So far, our characterisation of collaboration has focussed on aspects that are *internal* to specific design meetings—the processes of elaboration of mutual understanding and knowledge that they involve, as well as emergence of roles, mutual regulation of affect, and so on. But of course, situational factors greatly influence the way that collaboration proceeds: it is necessary to establish close links between processes that are internal to collaborative encounters, as well as what is 'outside' or around them, in the past (the history of the company, and previous collaborations of a group) and elsewhere in physical and social space (the project manager, client, etc.).

Engeström and Toivainen (2011) thus criticise approaches to collaboration that narrow them down to 'the analysis to the here and now, in other words, to very short slices of interaction with no history and no future'. (p. 35). They argue that, whilst 'such fine- grained analyses of discourse may be instructive, due to their isolation from more encompassing processes and historical changes, they have little to offer for researchers interested in problem- solving and learning embedded in complex transformations and design efforts typical to workplaces and professional networks'. (ibid., p. 35).

What is required, in order to situate collaborative encounters in their cultural, historical and socio-institutional settings, is to consider "what are the dominant patterns of organising collaboration in this activity at present, what were they in its preceding historical forms, and what might they become in the future?" (ibid., p. 35). There is a link here with particular 'cultures' of collaboration, in particular institutions and *milieux* as well as, indeed, Cultures. Collaboration is thus seen as a *social practice*, which has a developmental past, present and future; as an activity inscribed in particular social rules, a community of practice and division of labour.

As with any activity, collaborations have *objects* and *motives*; and these will influence both the way that collaboration proceeds as well as social actors' perceptions of its quality or qualities. If, for example, in a particular institution at a particular point in time, what is primary is to increase the profit of the company in which design activity takes place, to position it in the marketplace with a particular 'image', to innovate, to enable a new team to learn to work together, to induce new members, or any combination of these motives, then it is clear that these stakes will weigh upon the collaboration, and will constitute the primary criteria for the evaluation of its quality.

If collaborative design activities are situated within and across cultures, this means that—in addition to language, rituals and norms—*values* must play an important role in them. Values can be seen as rather stable and general principles underlying behaviour and judgement, that are seen as fundamental or important to people. Exactly what the generally accepted values that characterise a given society or culture is of course a matter for debate, within that culture and for researchers studying it. Nevertheless, social psychologists such as Schwartz and colleagues (e.g. Schwartz & Bilsky, 1987; Schwartz, 1994) have developed typologies of values, based on general factors such as openness to change *vs.* conservation, or self-transcendence *vs.* self-enhancement, and studied the extent to which specific values distinguish different cultures. For example, it has been claimed that Eastern cultures are more collectivist and Western ones individualist (Hofstede, 2001).

In collaborative design situations, values are primarily expressed in the domain of the *axiological*, that is concerning *judgements* as to what solution should (not) be accepted, rather than facts (Golder, 1996). But not all judgements involve values. For example, judgements could involve aesthetic or even economic hierarchies of preferences. We prefer to restrict value judgements to the domain of ethics, that is, to what is right or wrong.

Within collaborative creative design, values intervene in different ways: (a) according to the choice of what might be termed a general design paradigm (e.g. participatory design); (b) within the collaborative design process itself (expression of judgements of value relating to different choices), and (c) with respect to the values inherent in the design product (Détienne et al., 2019). The famous case of the first version of Google Glasses (augmented reality glasses that enabled people to make video recordings of their visual fields) illustrates the role of values in a design product in that, in certain cultures, the use of this artefact came into conflict with values such as the right to privacy. It is of course possible that cultures with their technologies could evolve such that an artefact that was hitherto unacceptable becomes banal.

We illustrate the roles of values and culture in the collaborative design process by the following two examples of specific research projects.

The first involved studying, appraising and comparing the quality collaborative engineering design processes, with respect to creative and innovative projects, across two institutional and national cultures: Télécom Paris, in France and Tokyo Institute of Technology, in Japan (Détienne et al., 2017). In each case (see the section on Appraising Collaboration Quality, below), students from both France and Japan appraised the quality of collaboration in videos of group work in their own culture and that of the other culture, using a specifically adapted 'Quality of Collaboration' evaluation method (Détienne et al., 2012a, 2012b). Across cultures/institutions, two main results emerged with respect to students' judgements of the quality of collaboration. Firstly, there was a strong common cross-cultural judgement with respect to aspects such as quality of problem solutions proposed, that related specifically to the domain of engineering. Secondly, there were significant differences in appraisals concerning aspects of the social dimension of interactions in groups, specifically with respect to the quality of collaborative resolutions of disagreements by argumentation processes. In brief, there was an almost complete absence of overt disagreement and argumentation in the Japanese groups, who preferred to preserve group harmony (*wa*). This is an important aspect to be taken into account in the implementation of group creativity workshops based on argumentation (see the discussion above) across different cultures.

The second research project that we evoke here raises the crucial issue as to how conflicts and disagreements concerning values can be resolved in collaborative design situations. Two types of research were carried out in the domain of collaborative design and functioning of shared housing projects (Le Bail et al., 2022). The first involved an in situ field study of a group project for creation of a new shared housing project. It was found that the mode of functioning of the nascent housing project was itself impregnated by values of mutual respect and the preservation of group harmony. Any proposal concerning what aspects were to be shared, and how new members were to be inducted or else excluded, avoided voting (that could split the collective into opposed subgroups) and preferred modifying rules and proposals to achieve complete consensus. A second research approach involved studying groups of four participants, in the laboratory, who were asked to co-design an ideal shared housing project in which they would agree to live. The participants were chosen according to their different types of academic training (business school, engineering school, design school, social sciences). If, as we stated above, values represent largely stable principles underlying action, to which people attach importance, then how can conflicts of values be resolved? In the lab study it was found that for some participants, the very idea of a collective housing project involved the (positive) value of 'sharing' resources, collaborating with respect to as many aspects as possible (see Andriessen & Baker, 2020, on collaboration as a positive value in itself). For others, however, they preferred to share as little as possible, stating that shared housing was not about values, but simply a means to save money. Overall, value-based disagreements were 'dissolved' rather than resolved, by simply shifting the focus of discussion to more factual questions, such as which form of (shared) heating would be most economical and ecological.

In sum, values permeate all levels of creative collaborative design, from the general principles concerning how it is to be organised, to actual social interactions in design situations and the way that values will facilitate or hinder the appropriation in society of the designed artefact. Taking value systems into consideration becomes particularly salient in situations involving cross-cultural design teams and where a co-design approach developed in one culture is to be transferred to another.

Appraising Collaboration: The Quality of Collaboration (QC) Method

Appraising collaboration requires capturing the dynamic and multi-level nature of the collaborative process. Unfortunately, this is not as easily quantifiable as group/team inputs and outputs such as task performance, satisfaction and changes at group/team and individual levels (Thornhill-Miller et al., 2023). There are indeed multiple interactions between the context, the collaboration processes, the task processes, and their (various) outcomes (Détienne et al., 2012b). Nevertheless, understanding what makes effective collaboration is of crucial importance for professional practice and training (Détienne et al., 2012a; Graesser et al., 2018), as evidenced by the long line of research on group or team collaboration over the past decades (for reviews, see, e.g., Salas et al., 2004; Mathieu et al., 2017).

The assessment of collaboration in professional situations has given rise to various methods, for example based on coding schemes applied to observations of meetings (e.g. Darses et al., 2004), focus-group to explore factors related to the quality of the 'subjective experiences' of collaboration (e.g. Van Dongen et al., 2016) or self-reported scales and questionnaires (e.g. Schroeder et al., 2001).

The quality of mediated collaboration has been assessed through two types of evaluation methods in Computer-Supported Cooperative Work (CSCW): usability inspection methods and user studies. Less costly than field methods, inspection methods focus on eliciting the goals and actions required for users to collaborate, and thus can be used earlier in the development process. Subsequently, they do not explore effective collaboration processes in context. User studies use a variety of data collection and analysis techniques (e.g. computer logs, interactions between participants, verbal communication of content) to investigate usability, adoption and usage patterns in organisations. Whether conducted in the field or in the laboratory, they tend to use few indicators to assess collaborative processes, focusing on quantifying fine-grained interactions (e.g. measures of 'communication effort' such as: number of speaker turns; number of words spoken; number of interruptions; number of basic questions asked). This approach has several limitations. Firstly, because these categories of analysis are often ad hoc with respect to the considered situation, they are difficult to apply in all situations and make it difficult to compare between studies. Secondly, quantitative variations of most of these indicators are non-univocal: any increase or decrease of them could signify either an interactive–intensive collaboration or else evidence of major difficulties in establishing and/or maintaining the collaboration. Thirdly, they usually concentrate only on one or two dimensions among the numerous ones to cover to assess the multiple dimensions of collaboration: for example, only examining the verbal and gestural communication to assess the grounding processes. Fourthly, motivational aspects as well as the balance/symmetry of individual contributions are not taken into account (Détienne et al., 2012b).

In the field of Computer-Supported Collaborative Learning (CSCL), the analysis of the process of collaboration has been a central topic of research. In particular, the rating scheme by Spada et al. (2005) is certainly the most representative of effort made in this field to compare and assess the dimensions of collaboration in collaborative learning tasks, with respect to various learning methods or technical support. We modified this approach so as to take into account the characteristics of activity in collaborative design situations (Burkhardt et al., 2009a, 2009b; Détienne et al., 2012b) and proposed a multi-dimensional rating scheme for evaluating the quality of collaboration in technology-mediated design. This scheme distinguishes seven dimensions, along five aspects, identified as central for collaboration in design: communication (fluidity of communication, sustaining mutual understanding), task processes (information exchanges for problem solving, argumentation and reaching consensus), group processes (task and time management, as well as symmetry in interaction, and individual task orientation (see Table 2.1).

The method relies on the *judgement* of an external observer(s) on an ongoing collaborative design process. The coding scheme consists in replying to 'yes/intermediate/no' questions on the different dimensions, leading to separate dimension scores and a global score of 'quality of collaboration'. The obtained score is not an absolute score of 'ideal' collaboration, but is a powerful tool for contrasting groups and situations in comparable contexts.

This method has been used to analyse the quality of collaboration of two student teams in a longitudinal study of a collaborative distant architectural studio (Safin et al., 2019). The results highlight that the quality of

Table 2.1 Dimensions of collaboration in relation to appraisal of its quality (Détienne et al., 2012b)

Dimensions	*Definition*	*Indicators*
1. Fluidity of collaboration	It assesses the management of verbal communication (verbal turns), of actions (tool use) and of attention orientation	– Fluidity of verbal turns – Fluidity of tools use (stylet, menu) – Coherency of attention orientation
2. Sustaining mutual understanding	It assesses the grounding processes concerning the design artefact (problem, solutions), the designers' actions and the state of involved tools	– Mutual understanding of the state of design problem/solutions – Mutual understanding of the actions in progress and next actions – Mutual understanding of the state of the system (active functions, open documents)
3. Information exchanges for problem solving	It assesses design ideas pooling, refinement of design ideas and coherency of ideas	– Generation of design ideas (problem, solutions, past cases, constraints) – Refinement of design ideas – Coherency and follow up of ideas
4. Argumentation and reaching consensus	It assesses whether or not there is argumentation and decision taken on common consensus	– Criticisms and argumentation – Checking solutions adequacy with design constraints – Common decision taking
5. Task and time management	It assesses the planning (e.g. task allocation) and time management	– Work planning – Task division – Distribution and management of tasks interdependencies – Time management

(*continued*)

Table 2.1 (continued)

Dimensions	*Definition*	*Indicators*
6. Cooperative orientation	It assesses the balance of contribution of the actors in design, planning, and in verbal and graphical actions	– Symmetry of verbal contributions – Symmetry of use of graphical tools – Symmetry in task management – Symmetry in design choices
7. Individual task orientation	It assesses, for each contributor, its motivation (marks of interest in the collaboration), implication (actions) and involvement (attention orientation)	– Showing up motivation and encouraging others motivation – Constancy of effort put in the task – Attention orientation in relation with the design task

collaboration and the design process appear to have a bilateral relationship: good collaboration allows the design process to progress, and the progression of the design process provides the conditions for good collaboration.

Beyond the assessment of collaborative design processes, this approach has been used to assess several professional collaborative activities, such as collaboration between actors in crisis management (Gaudin et al., 2011a, 2011b), patient-therapist collaboration (Wrzesien et al., 2012), criminal evidence collection (Poelman et al., 2012), or the dynamic representation of 'personas' (Bonnardel & Pichot, 2020). More recently, it has been extended to support cross-cultural comparisons of collaborative design processes (Détienne et al., 2017).

Collaboration and Its Quality Must Be Understood from Multiple Perspectives

The very idea of 'quality of collaboration' might appear, at first sight, to be reduced to that of researchers' normative perspectives on what constitutes good or effective collaboration, as a process, with respect to outcomes, and the relation between the two. Such a perspective exists and is, we would claim, legitimate (cf. Spada et al., 2005; Détienne et al., 2008). But it is only one perspective amongst many. This 'researcher' perspective is often part of a research paradigm whereby qualities of the collaboration process are considered as intermediary variables, to be correlated with its outcomes (learning, quality of product).

Whilst speaking of quality of collaboration is in itself inherently a judgement, this can and must nevertheless be considered from multiple perspectives, of participants and other social actors involved in collaborative design, in relation to multiple stakes and objects of activity, in order to ensure social and practical relevance of the appraisal.

Participants in design have of course their own experiences and perspectives of collaboration. In relation to, yet also relatively independent from outcomes, collaboration can be a specific socio-relational personal *experience* that may be more or less personally rewarding (Crook, 1994). In multi-expertise situations, participants may feel that their own point of view has been more or less taken into account.

Whilst participants will be personally concerned with the quality of the collaboration process, as well as its products, it is probable that managers and clients will be more concerned with the quality of its outcomes, perhaps irrespectively of how these are achieved. An exception is the case where a design meeting has also an orientation towards learning to collaborate together, or 'team building'.

This highlights three points. Firstly, the evaluation of the quality of collaboration must be seen from the points of view of different engaged social actors; secondly, depending on these perspectives, different aspects of collaboration will be emphasised; thirdly, emphasis on process or product, constructive or productive interaction, and the relations between, the two, will be situated in the articulation between, perspectives of participants present or absent in the design meeting. Participants may of course care just as much about the quality of products of design as external persons (such as owners, managers, clients), and it is of course possible that managers may care about the quality of collaboration processes, to the extent that they consider broader perspectives and temporal scales, relating to general development of collaborative teams, of individuals' skills in design, and the future of the company. If managers are able to take a longer term view of the degree of social cohesion in their enterprises, they may well realise that quality of products is not necessarily the only factor, especially if it requires a collaboration process that is demotivating for participants (e.g. the cost in new personnel recruitment, absenteeism, lack of implication in the company's shared narrative, etc. may therefore be higher).

There is thus no 'neutral' or 'God's eye' view of the quality or qualities of collaboration, in the case of design, or in general: normative-descriptive methods for its analysis are necessarily from a particular perspective, relating to specific goals, stakes and objects of activity.

Concluding Discussion

This chapter synthesises dimensions of collaboration that have been described in the literature, and in particular with respect to creative design situations. Collaboration is a particular kind of collective activity, with a high degree of interdependence between actors, engaged together in a common task with a common goal and a shared focus. It is differentiated from other forms of collective activity such as coordination or cooperation, notably by a relative equity between actors, related to their rights to contribute. When designing collaboratively, people engage in different activities: task-related activities, such as expressing ideas and evaluating proposals, but also process-oriented activities such as coordination, group organisation and regulation behaviours, that may be more or less guided by scripts and procedures. In this regard, external representations and design instruments play several roles, as object of the design activity, support for the task evolution, and as coordination tools. Taken as an ensemble, collaborative design involves several types of processes: discursive (argumentation, explanation, elaboration), socio-relational (group dynamics), affective (emotional), creative (oriented towards novelty), constructive (collaboration builds itself and evolves through time), productive (task-oriented) and reflexive (the process in itself is an object of thinking). These processes are intertwined, and are grounded in values in a cultural context, which impacts the product, but also the collaboration activity in itself.

For these reasons, we propose that appraising collaborative design requires a multidimensional approach, allowing the diversity of processes and influential factors to be taken into account. We introduced the QC (Quality of Collaboration) evaluation method, which allows rating of the quality of collaboration along several dimensions, based on a judgemental coding scheme, and comparing situations to one another.

There is still much to be learned about the collaborative processes of design and creativity and there are research avenues still to be deeply explored. We mention three of them here.

The first one concerns how to articulate viewpoints, in particular, the first-person perspectives of the participants with the third-party perspective of the researcher. Depending on the objective of research, be it transformative (meaning enhancing the design process) and/or evaluative/reflexive (analysing collaboration), this question opens clear directions for future research. Auto-confrontation techniques, allowing participants to review and comment on their own processes, are powerful tools to be combined with more classical coding schemes used by researchers to characterise activities.

This naturally leads to the second topic, about the possibilities offered by AI (Artificial Intelligence), sensors and IoT, enabling to automatically process and analyse behavioural data (e.g. Cukurova et al., 2017). Such automated (mostly behavioural) data analytics could be put in perspective with more sensitive, subjective data obtained by first-person view methods, as well as deep theories of collaboration.

Finally, there is still a need to gather more knowledge about the developmental processes of collaboration, that should be studied in longitudinal complex ecological settings: how organisations develop a 'culture of creative collaboration', how emerging technologies empower or constraint design teams, how individual get more skilled on the collaborative processes in professional design contexts.

References

Akin, Ö., & Akin, C. (1998). On the process of creativity in puzzles, inventions, and designs. *Automation in Construction, 7*, 123–138.

Allwood, J., Traum, D., & Jokinen, K. (2000). Cooperation, dialogue and ethics. *International Journal of Human-Computer Studies, 53*(6), 871–914.

Andriessen, J., & Baker, M. J. (2020). *On collaboration: Personal, educational and societal arenas.* Brill|Sense Publishers. https://doi.org/10.1163/9789004429086

Andriessen, J., Baker, M. J., & Suthers, D. (2003). Arguing to learn: Confronting cognitions in computer-supported collaborative learning environments. Kluwer. https://doi.org/10.1007/978-94-017-0781-7

Andriessen, J., Baker, M., & van der Puil, C. (2011). Socio-cognitive tension in collaborative working relations. In S. Ludvigsen, A. Lund, I. Rasmussen, &

R. Saljo (Eds.), *Learning across sites: New tools, infrastructures and practices* (pp. 222–242). Routledge.
Badke-Schaub, P. (1999). Group effectiveness in design practice: Analysis and training by a critical-situation-approach. *Psychologische Beiträge, l*(41), 338–355.
Badke-Schaub, P., & Buerschaper, C. (2001). Creativity and complex problem solving in the social context. In C. M. Allwood & M. Selart (Eds.), *Decision making: Social and creative dimensions* (pp. 177–196). Kluwer.
Badke-Schaub, P., Goldschmidt, & Meijer, M. (2010). How does cognitive conflict in design teams support the development of creative ideas? *Creativity and Innovation Management, 19*(2), 119–133.
Baker, M. (1994). A model for negotiation in teaching-learning dialogues. *Journal of Artificial Intelligence in Education, 5*(2), 199–254.
Baker, M. J. (1999). Argumentation and constructive interaction. In G. Rijlaarsdam & E. Espéret (Series Eds.) and P. Coirier & J. Andriessen (Vol. Eds.). *Studies in writing: Vol. 5. Foundations of argumentative text processing* (179–202). University of Amsterdam Press.
Baker, M. (2002). Forms of cooperation in dyadic problem-solving. *Revue d'Intelligence Artificielle, 16*(4–5), 587–620.
Baker, M. J. (2015). Collaboration in collaborative learning. *Interaction Studies: Social Behaviour and Communication in Biological and Artificial Systems, 16*(3), 451–473. https://doi.org/10.1075/is.16.3.05bak
Baker, M., Andriessen, J., & Järvelä, S. (Dir.) (2013). *Affective learning together: Social and emotional dimensions of collaborative learning*. Routledge.
Baker, M. J., Détienne, Lund, K., & Séjourné, A. (2009). Étude des profils interactifs dans une situation de conception collective en architecture. In F. Détienne & V. Traverso (Eds.), *Méthodologies d'analyse de situations coopératives de conception: Corpus MOSAIC* (pp. 183–220). Presses Universitaires de Nancy.
Baker, M. J., Détienne, F., Mougenot, C., Corvin, T., & Pennington, M. (2020). Argumentation, eureka and emotion: An analysis of group projects in creative design training. *Learning, Culture and Social Interaction, 26*, 100436. https://doi.org/10.1016/j.lcsi.2020.100436
Baker, M., & Lund, K. (1997). Promoting reflective interactions in a CSCL environment. *Journal of Computer Assisted Learning, 13*, 175–193.
Bales, R. F. (1950). A set of categories for the analysis of small group interaction. *American Sociological Review, 15*, 257–263.
Barcellini, F., Détienne, F., & Burkhardt, J. M. (2014). A situated approach of roles and participation in Open Source Software Communities. *Human Computer Interaction, 29*(3), 205–255.
Barth, E. M., & Krabbe, E. C. W. (1982). *From axiom to dialogue: A philosophical study of logics and argumentation*. Walter de Gruyter.

Beers, P., Boshuizen, H., Kirschner, P., & Gijselaers, W. (2006). Common ground, complex problems and decision making. *Group Decision and Negotiation, 15*, 529–556.

Bonnardel, N. (2006). *Créativité et conception. Approches cognitives et ergonomiques.* Solal.

Bonnardel, N. (2009). Activités de conception et créativité: de l'analyse des facteurs cognitifs à l'assistance aux activités de conception créatives. *Le Travail Humain, 72*(*1*), 5–22.

Bonnardel, N., & Pichot, N. (2020). Enhancing collaborative creativity with virtual dynamic personas. *Applied Ergonomics, 82*, 102949.

Brown, P., & Levinson, S. (1987). *Politeness: Some universals in language usage.* Cambridge University Press.

Bucciarelli, L. L. (1988). An ethnographic perspective on engineering design. *Design Studies, 9*(*3*), 159–168.

Burkhardt, J.-M., Détienne, F., Hébert, A.-M., & Perron, L. (2009a). Assessing the "Quality of Collaboration" in technology-mediated design situations with several dimensions. In T. Gross, J. Gulliksen, P. Kotzé, L. Oestreicher, P. Palanque, R. O. Prates, & M. Winckler (Eds.), *Human-computer interaction – INTERACT 2009* (pp. 157–160). Springer.

Burkhardt, J.-M., Détienne, F., Hebert, A.-M., Perron, L., Safin, S., & Leclercq, P. (2009b, September 30–October 2). An approach to assess the quality of collaboration in technology-mediated design situations. In *Proceedings of the European conference on cognitive ergonomics ECCE 2009.*

Clark, H. H., & Brennan, S. E. (1991). Grounding in communication. In L. Resnick, J.-M. Levine, & S. D. Teasley (Eds.), *Perspectives on socially shared cognition* (pp. 127–149). APA.

Clark, H. H., & Schaefer, E. F. (1989). Contributing to discourse. *Cognitive Science, 13*, 259–294.

Crook, C. (1994). *Computers and the collaborative experience of learning.* Routledge.

Cukurova, M., Luckin, R., Mavrikis, M., & Millán, E. (2017). Machine and human observable differences in groups' collaborative problem-solving behaviours. In É. Lavoué, H. Drachsler, K. Verbert, J. Broisin, & M. Pérez-Sanagustín (Eds.), *Data driven approaches in digital education* (EC-TEL 2017. Lecture notes in computer science) (Vol. 10474). Springer. https://doi.org/10.1007/978-3-319-66610-5_2

Darses, F., Détienne, F., & Visser, W. (2004). Les activités de conception et leur assistance. In P. Falzon (Ed.), *Ergonomie.* PUF.

Détienne, F., Baker, M., & Burkhardt, J.-M. (2012a). Perspectives on quality of collaboration in design. *CoDesign, 8*(*4*), 197–199.

Détienne, F., Baker, M., & Burkhardt, J.-M. (2012b). Quality of collaboration in design meetings: Methodological reflexions. *CoDesign: International Journal of CoCreation in Design*, *8*(4), 247–261. https://doi.org/10.1080/15710882.2012.729063

Détienne, F., Baker, M., Fréard, D., Barcellini, F., Denis, A., & Quignard, M. (2016). The Descent of Pluto: Interactive dynamics, specialisation and reciprocity of roles in a Wikipedia debate. *International Journal of Human-Computer Studies, Vol.*, *86*(February 2016), 11–31. https://doi.org/10.1016/j.ijhcs.2015.09.002

Détienne, F., Baker, M., & Le Bail, C. (2019). Ideologically-embedded design: Community, collaboration and artefact. *International Journal of Human-Computer Studies, 131*, 72–80. https://doi.org/10.1016/j.ijhcs.2019.06.003

Détienne, F., Baker, M., Vanhille, M., & Mougenot, C. (2017). Cultures of collaboration in engineering design education: A contrastive case study in France and Japan. *International Journal of Design Creativity and Innovation*, *5*(1-2), 104–128. https://doi.org/10.1080/21650349.2016.1218796

Détienne, F., Burkhardt, J.-M., Hébert, A.-M., & Perron, L. (2008, November 9). Assessing the quality of collaboration in design: Bridging cognitive ergonomics and CSCL approaches. In *Workshop « CSCW and Human Factors », CSCW'2008.*

Détienne, F., Martin, G., & Lavigne, E. (2005). Viewpoints in co-design: A field study in concurrent engineering. *Design Studies, 26*, 215–241.

Dillenbourg, P. (2002). Over-scripting CSCL: The risks of blending collaborative learning with instructional design. In P. A. Kirschner (Ed.), *Three worlds of CSCL. Can we support CSCL?* (pp. 61–91). Open Universiteit Nederland.

Engeström, Y., & Toivainen, H. (2011). Co-configurational design of learning instrumentalities: An activity-theoretical perspective. In S. Ludvigsen, A. Lund, I. Rasmussen, and R. Säljö (Eds.), *Learning Across Sites: New tools, infrastructures and practices* (pp. 33–52). Routledge.

Engeström, Y., Brown, K., Christopher, L. C., & Gregory, J. (1997). Coordination, cooperation, and communication in the courts: Expansive transitions in legal work. In M. Cole, Y. Engeström, & O. Vasquez (Eds.), *Mind, culture and activity* (pp. 369–385). Cambridge University Press.

Falzon, P. (1994). Dialogues fonctionnels et activité collective. *Le Travail Humain, 57*, 297–312.

Fischer, F., Kollar, I., Mandl, H., & Haake, J. M. (2007). *Scripting computer-supported collaborative learning: Cognitive, computational and educational perspectives.* Springer.

Fréard, D., Denis, A., Détienne, F., Baker, M., Quignard, M., & Barcellini, F. (2010, August 25–27). The role of argumentation in online epistemic communities: the anatomy of a conflict in Wikipedia. In *Proceedings of the European conference on cognitive ergonomics ECCE 2010.*

Gaudin, C., Delgoulet, C., Gounelle, C., Verneuil, L., & Burkhardt, J. M. (2011a). Évaluation de la qualité de la collaboration lors d'une situation à risque: le cas

de la gestion d'un événement NRBC par une équipe multidisciplinaire. In *46e congrès de la Société d'Ergonomie de Langue Française* (pp. 222–228).
Gaudin, C., Philibert, B., & Bonnardel, N. (2011b, septembre 5–7). Effet de la qualité de la collaboration sur la performance lors de la résolution de problèmes en équipe. In *Actes des Journées de Psychologie Ergonomique EPIQUE 2011.*
Golder, C. (1996). *Le développement des discours argumentatifs.* Delachaux & Niestlé.
Graesser, A. C., Fiore, S. M., Greiff, S., Andrews-Todd, J., Foltz, P. W., & Hesse, F. W. (2018). Advancing the science of collaborative problem solving. *Psychological Science in the Public Interest, 19*, 59–92.
Heiner, J., Tversky, B., & Silverman, M. (2004). Sketches for and from collaboration. In *Paper presented at the visual and spatial reasoning in design III.* Key Centre for Design Research.
Hofstede, G. (2001). *Culture's consequences: Comparing values, behaviors, institutions, and organizations across nations* (2nd ed.). Sage Publications.
Johnson, D. W., & Johnson, F. (1994). *Joining together: Group theory and group skills* (5th ed.). Prentice-Hall.
Kristensson, P., Gustafsson, A., & Archer, T. (2004). Harnessing the creative potential among users. *The Journal of Product Innovation Management, 21*(*1*), 4–14.
Le Bail, C., Baker, M., & Détienne, F. (2022). Values and argumentation in collaborative design. *CoDesign, International Journal of CoCreation in Design and the Arts, 18*(*2*), 165–185.
Leont'ev, A. N. (1979). The problem of activity in psychology. In J. V. Wertsch (Ed.), *The concept of activity in Soviet psychology.* Sharpe, Inc.
Leplat, J. (2004). L'analyse psychologique du travail. *Revue Européenne de Psychologie Appliquée., 54*(*2*), 101–108.
Loup-Escande, E., Burkhardt, J. M., Christmann, O., & Richir, S. (2014). Needs' elaboration between users, designers and project leaders: Analysis of a design process of a virtual reality-based software. *Information and Software Technology, 56*(*8*), 1049–1061.
Lubart, T. I. (1994). Creativity. In R. J. Sternberg (Ed.), *Thinking and problem solving* (pp. 289–332). Academic Press.
Mann, W. C., & Thompson, S. A. (1988). Rhetorical Structure Theory: Toward a functional theory of text organisation. *Text, 8*(*3*), 243–281.
Mathieu, J. E., Hollenbeck, J. R., van Knippenberg, D., & Ilgen, D. R. (2017). A century of work teams in the Journal of Applied Psychology. *The Journal of Applied Psychology, 102*, 452–467.
Mer, S., Jeantet, C., & Tichkiewitch, S. (1995). Les objets intermédiaires de la conception: modélisation et communication. In J. Dans Caelen & K. Zreik (Eds.), *Le Communicationnel pour Concevoir* (pp. 21–41). Europia Productions.

Miyake, N. (1986). Constructive interaction and the iterative process of understanding. *Cognitive Science, 10*, 151–177.
Muntigl, P., & Turnbull, W. (1998). Conversational structure and facework in arguing. *Journal of Pragmatics, 29*, 225–256.
Poelman, R., Akman, O., Lukosch, S., & Jonker, P. (2012). As if being there: Mediated reality for crime scene investigation. In *Proceedings of the ACM 2012 conference on computer supported cooperative work* (pp. 1267–1276).
Richard, P. (2016). *La participation des usagers au processus de conception créative de solutions de mobilité. Pratiques, impact et préconisations.* PhD Dissertation, Université Paris 8.
Rietveld, E., & Brouwers, A. A. (2017). Optimal grip on affordances in architectural design practices: An ethnography. *Phenomenology and the Cognitive Sciences, 16*, 545–564.
Roschelle, J., & Teasley, S. D. (1995). The construction of shared knowledge in collaborative problem solving. In C. E. O'Malley (Ed.), *Computer-supported collaborative learning* (pp. 69–197). Springer-Verlag.
Safin, S. (2025). Les représentations graphiques en conception et participation. In S. Safin (Ed.), *Les activités cognitives de conception en architecture.* ISTE: Collection Sciences.
Safin, S., Détienne, F., Burkhardt, J.-M., Hébert, A.-M., & Leclercq, P. (2019). The interplay between quality of collaboration, design project evolution and outcome in an architectural design studio. *CoDesign, 17*(*4*), 392–409.
Safin, S., Dorta, T., & Kinayoglu, G. (2018, October 23–26). Patterns of creative UX in immersive collaborative design. In *Proceedings of IHM 2018 - 30e Conférence Francophone sur l'Interaction Homme-Machine.*
Safin, S., Juchmes, R., & Leclercq, P. (2011). Du crayon au stylo numérique: influences des IHM à stylo et des interprétations numériques sur l'activité graphique en tâches de conception. *Journal d'Interaction Personne-Système, 2*(*1*).
Safin, S., Juchmes, R., & Leclercq, P. (2012). Use of graphical modality in a collaborative design distant setting. Dans *Proceedings of COOP 2012.*
Salas, E., Stagl, K. C., & Burke, C. S. (2004). 25 years of team effectiveness in organizations: Research themes and emerging needs. In *International review of industrial and organizational psychology* (pp. 47–91). John Wiley & Sons, Ltd.
Schmidt, K. (1994). Cooperative work and its articulation: Requirements for computer support. *Le Travail Humain, 57*(*4*), 345–366.
Schön, D. A. (1983). *The reflective practitioner: How professionals think in action.* Basic Books.
Schroeder, R., Steed, A., Axelsson, A. S., Heldal, I., Abelin, Å., Wideström, J., et al. (2001). Collaborating in networked immersive spaces: As good as being there together? *Computers & Graphics, 25*(*5*), 781–788.
Schwartz, S. H. (1994). Are there universal aspects in the structure and contents of human values. *Journal of Social Issues, 50*(4), 19–45.

Schwartz, S. H., & Bilsky, W. (1987). Toward a universal psychological structure of human values. *Journal of Personality and Social Psychology, 53*(3), 550–562.

Simon, H. A. (1995). Problem forming, problem finding, and problem solving in design. In A. Collen & W. W. Gasparski (Eds.), *Design and systems: General applications of methodology*. Transaction Publishers.

Spada, H., Meier, A., Rummel, N., & Hauser, S. (2005). A new method to assess the quality of collaborative process in CSCL. In T. Koschmann, D. Suthers, & T. W. Chan (Eds.), *Proceedings of computer supported collaborative learning 2005: The next 10 years!* (pp. 622–631). Lawrence Erlbaum Associates.

Star, S. L., & Griesemer, J. R. (1989). Institutional ecology, 'Translations' and boundary objects: Amateurs and professionals in Berkeley's museum of vertebrate zoology, 1907–39. *Social Studies of Science, 19*(3), 387–420. https://doi.org/10.1177/030631289019003001

Suwa, M., Gero, J., & Purcell, T. (2000). Unexpected discoveries and S-invention of design requirements: Important vehicles for a design process. *Design Studies, 21*, 539–567.

Thornhill-Miller, B., Camarda, A., Mercier, M., Burkhardt, J. M., Morisseau, T., Bourgeois-Bougrine, S., et al. (2023). Creativity, critical thinking, communication, and collaboration: Assessment, certification, and promotion of 21st century skills for the future of work and education. *Journal of Intelligence, 11*(*3*), 54.

Valkenburg, R. C. (2000). *The reflective practice in product design teams Industrial Engineering Design*. Delft University of Technology.

Van den Bossche, P., Gijselaers, W. H., Segers, M., & Kirschner, P. A. (2006). Social and cognitive factors driving teamwork in collaborative learning environments: Team learning beliefs and behaviors. *Small Group Research, 37*(5), 490–521. https://doi.org/10.1177/1046496406292938

Van Dongen, J. J. J., Lenzen, S. A., van Bokhoven, M. A., Daniëls, R., van der Weijden, T., & Beurskens, A. (2016). Interprofessional collaboration regarding patients' care plans in primary care: A focus group study into influential factors. *BMC Family Practice, 17*, 1–10.

Vignaux, G. (1990). Chapter 40: A cognitive model of argumentation. In F. H. van Eemeren, R. Grootendorst, J. A. Blair, & C. A. Willard (Eds.), *Proceedings of the second international conference on argumentation* (Vol. 1, pp. 303–310).

Visser, W. (2006). *The cognitive artifacts of designing*. Lawrence Erlbaum Associates.

Walton, D. N. (1992). *Plausible argument in everyday conversation*. State University of New York Press.

Wrzesien, M., Burkhardt, J. M., Botella, C., & Alcañiz, M. (2012). Evaluation of the quality of collaboration between the client and the therapist in phobia treatments. *Interacting with Computers, 24*(*6*), 461–471.

Yong, K., Sauer, S. J., & Mannix, E. A. (2014). Conflict and creativity in interdisciplinary teams. *Small Group Research, 45*(*3*), 266–289.

CHAPTER 3

The Psychosocial and Cognitive Underpinnings of Collective and Organizational Creativity

Fabien Girandola and Nathalie Bonnardel

Introduction

Social interactions and dynamics can be both a source of innovation and a support for creativity. The approach developed in this chapter draws on social, cognitive, and ergonomic psychology, depending on the aspects addressed, to analyze how exchanges and, in particular, group collaboration influence the emergence of creativity. In a context where creativity is

F. Girandola (✉)
Aix-Marseille Université, Laboratoire de Psychologie Sociale (LPS, UR 849), Aix-en-Provence, France

Institute for Creativity and Innovations of Aix-Marseille (InCIAM), Aix-en-Provence, France
e-mail: fabien.girandola@univ-amu.fr

N. Bonnardel
Institute for Creativity and Innovations of Aix-Marseille (InCIAM), Aix-en-Provence, France

Aix-Marseille Université, Centre de recherche en Psychologie de la Connaissance, du Langage et de l'Emotion (PSYCLE, UR 3273), Aix-en-Provence, France
e-mail: nathalie.bonnardel@univ-amu.fr

N. Bonnardel (ed.), *Collaborations in Creativity*, The Seven C's of Creativity, https://doi.org/10.1007/978-3-032-07753-0_3

most often a strategic lever for organizations (companies, public institutions, etc.), it becomes essential to study the underlying collaborative processes and those that can contribute to the smooth functioning of the collective (e.g., Bonnardel et al., 2023; Paulus & Nijstad, 2019; Pellegrin et al., 2021; Reiter-Palmon & Hunter, 2023a, 2023b; Rothouse, 2020). The fundamental question guiding our reflections in this chapter is: how do socio-cognitive dynamics influence creativity? Based on this chapter, certain factors contributing to a socio-cognitive environment that stimulates creativity will be highlighted.

Models of Creativity in Collective Situations

Amabile's Component Model (1983, 1998)

Amabile (1983, 1998) distinguished herself from approaches that emphasized the traits and profiles of creative personalities. She proposes a component model based on (1) intrinsic motivation, the cornerstone of the model, which relates to the personal interest of the individual (the person engaged in a creative activity), their own determination towards their work and the completion of the task (Amabile, 1997; Amabile & Michael, 2016). Thus, individuals "are intrinsically motivated when they seek pleasure, interest, satisfaction of their curiosity, self-expression, or self-actualization in their work" (Amabile, 1993, p. 188). The environment in which the individual evolves must allow this motivation to develop or be reinforced (e.g., goals, recognition, trust). This model also highlights (2) expertise or skills in the field (i.e., specialized skills and knowledge, meaning everything a person knows and can do in a field; e.g., Cromwell, 2020) as well as (3) creative skills. Several dimensions are measurable: flexibility of thinking, originality, the ability to combine existing ideas in new and interesting ways, and affect (Amabile et al., 1986, 2005). Creative skills include a cognitive style that promotes the integration and combination of new perspectives, the ability to take risks, and perseverance. Amabile also emphasized the importance of context and environment (e.g., support from management, management practices, resources, climate), particularly social and organizational, in fostering intrinsic motivation, which in turn influences creative processes (Amabile et al., 2004; Reiter-Palmon et al., 2020). It also appears important to foster a collaborative atmosphere and accept failure (Amabile, 1992, 2004; Amabile & Conti, 1999; Amabile & Sensabaugh, 1992; Amabile et al., 2002a, 2002b). For Amabile

(1998), organizational creativity "is the production of novel and useful ideas in any domain by individuals and teams, which is a starting point for innovation." Despite all their contributions to the understanding of creativity, Amabile's perspectives would benefit from being supplemented, particularly with regard to the consideration of social dynamics and interactions within groups. Thus, while recognizing the important contribution of Amabile's work, Paulus (2020) points to a lack of theorization on motivational synchrony in creative teams, in the development of hybrid methods, and in the alternation between solitary and interactive work according to creative phases. For Paulus (2020), research needs to be developed on the antagonistic factors that hinder collective creativity: social anxiety during interactions, negative comparisons between members, inappropriate distribution of roles and skills, and the gender composition of groups. Paulus proposes modeling these group mechanisms to explain creative failures (Paulus, 2000). These recommendations highlight the desire to enrich Amabile's model with a systemic approach that better integrates the social, cognitive, emotional, and structural dimensions of collective creative processes (Paulus & Dzindolet, 2008, see Figure 3).

The Interactionist Model of Woodman, Sawyer, and Griffin (1993)

This model accounts for the interaction between individual characteristics (cognitive abilities, personality, intrinsic motivation, knowledge), group characteristics (norms, cohesion, size, roles, tasks, diversity, problem-solving techniques), and organizational characteristics (culture, resources, rewards, strategy, structure, technology). These interactions influence creative processes, promoting the transformation of ideas into products or innovations (see Fig. 3.1). In short, this model highlights the complexity of social and organizational interactions that foster a context conducive to creative behavior (see Fig. 3.2). According to Woodman et al. (1993), "organizational creativity is the creation of a valuable, useful new product, service, idea, procedure, or process by individuals working together in a complex social system" (p. 293). Furthermore, according to Ford (1996), "organizational creativity as a specific domain is a subjective judgment of the novelty and value of an outcome of a particular action."

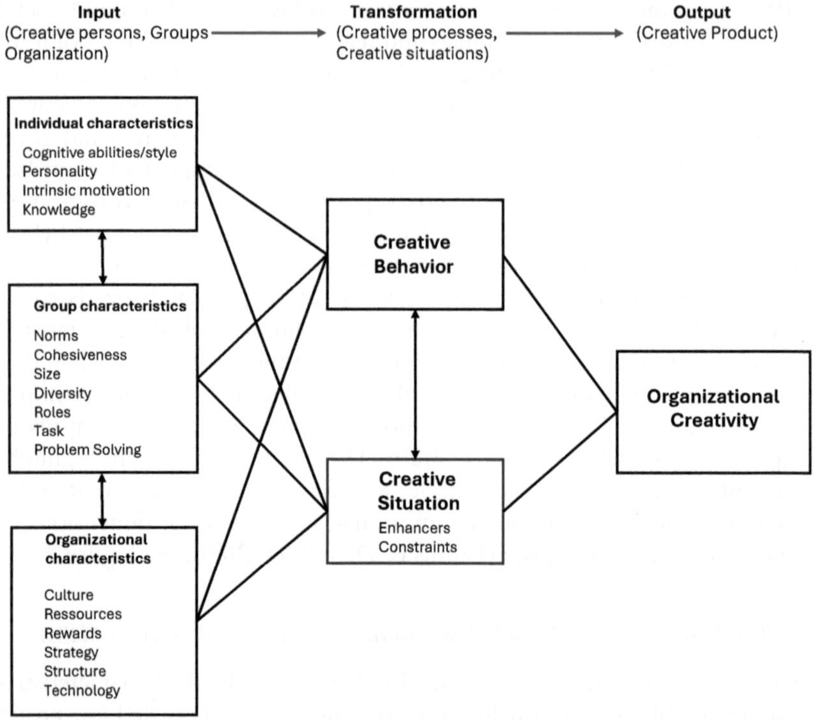

Fig. 3.1 Conceptual links among creative persons, processes, situations, and products (based on Woodman et al., 1993, p. 309)

Paulus and Dzindolet's Model of Collective Creativity and Innovation (2008)

Paulus and Dzindolet (2008) present a model emphasizing the interaction of social, cognitive, and motivational variables in situations of collective creativity (see Fig. 3.3). These variables can stimulate or inhibit the generation of creative ideas. They do not function in isolation but interact in a circular and dynamic manner. For example, creative outcomes in turn influence the processes underlying group creativity.

The models of Woodman et al. (1993) and Paulus and Dzindolet (2008) adopt a multilevel systemic perspective that takes into account individual, group, and organizational factors. The model developed by Woodman et al. (1993) places particular emphasis on organizational

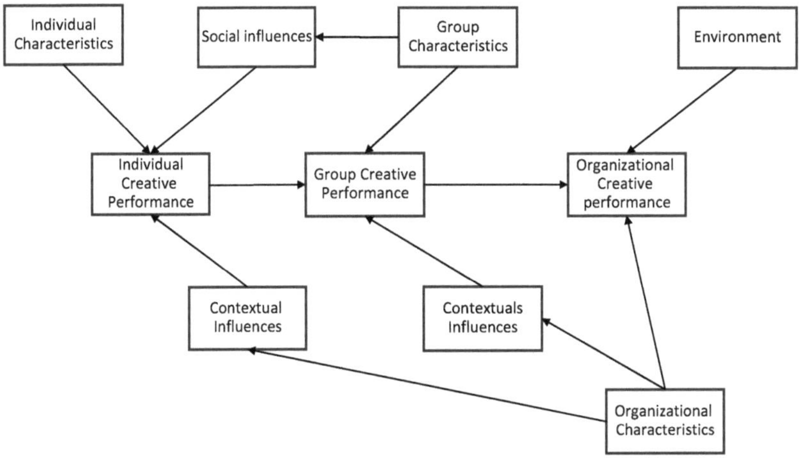

Fig. 3.2 Hypothesis linkages among factors related to organizational creativity (based on Woodman et al., 1993, p. 311)

factors (e.g., organizational culture, support for innovation, leadership) and interactions between organizational levels: individuals, groups, teams, and organizational culture (see also Dechamp & Szostak, 2016; Leonard-Barton & Swap, 1999). In the same vein, the model developed by Paulus and Dzindolet (2008) places a high emphasis on the collective, the role of social influences in groups and teams, and intra- and inter-group aspects (Paulus et al., 2021). The points discussed in the rest of this chapter will also highlight the complex interactions between the individual, social, and organizational dimensions that promote creativity (for further work on the analysis of cognitive and social aspects in collective creative activities, see Baker et al., Chap. 2 in this book, and Christenson et al., Chap. 4), while also pointing out some new developments in the paradigms involved in collective and organizational creativity (see Fillon et al., 2024). We will successively address psychosocial influences on creativity in collective situations, and creative teamwork associated with certain brainstorming techniques. Then, we will describe the impact of leadership on creativity, and conclude by highlighting various aspects related to organizational creativity.

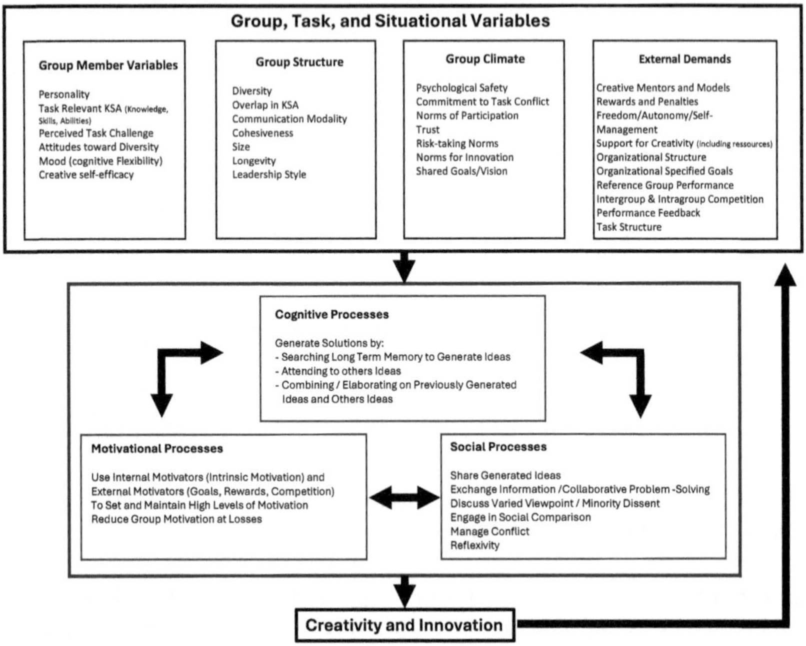

Fig. 3.3 A model of collective creativity and innovation (based on Paulus & Dzindolet, 2008, p. 230)

Psychosocial Influences on Collective Creativity

The Psychosocial Perspective and the Role of the Audience

Social psychology research shows that the presence of others has an effect on individual and collective performance (Moscovici, 1984). Individual positions are determined in and by the relationship with the social environment. The presence of others, even in a symbolic or imaginary way, influences an individual's performance, whether it is the person who assigned the task (e.g., the employer or project manager in a team) or other participants who are jointly performing the same task in co-action, that is, with reciprocal influence between members of the same team. These results can be supplemented by the work of Belletier and Camos (2018), who explore social influence on working memory by showing that social factors, such as the presence of an experimenter, can modulate cognitive performance.

The presence of others is also expressed with non-human agents. Thus, some research explores interactions between humans and robots and their effects on sociocognitive processes. Bouquet et al. (2024), for example, examine how individuals integrate their actions with those of human or robotic partners in collaborative tasks. Integration between oneself and others refers to the ability to coordinate one's own actions with those of a partner to achieve a common goal, as described in the COFOR model in Chap. 1 (by Bonnardel, Girandola, Lellouche-Gounon, Lopez-Bagousse & Toumi). Other research highlights the interaction between social and cognitive processes and their effects (see facilitation and social loafing, Belletier et al., 2019; Latané et al., 1979) in contexts involving human or artificial partners, as will be discussed in Chap. 8 (by Chompunuch, Vinchon & Lubart) and in Chap. 9 (by Bonnardel, Marcon, Moták & Belletier).

Furthermore, some authors have established a link between this psychosocial perspective and creativity. Haslam et al. (2015) demonstrated the importance of social consensus in the acceptance of ideas. Ideas are only recognized, appreciated, and used if they spread within a community and reach an audience. For them, social recognition is particularly decisive in the creative process. Thus, a person's creative behavior is likely to be influenced by the values and preferences of the group to which the creator belongs. Participants whose collective identity is emphasized judge ideas to be more creative when they fall within the norms of the reference group. When personal identity is emphasized, participants judge ideas that deviate from the norms of their reference or membership group as more creative (Adarves-Yorno et al., 2006; van Bezouw et al., 2021).

Conformity and Majority Influence

Asch (1951, 1955, 1956) shows that conformity results from an interaction between an individual and social pressure. "Conformity refers to a change in perception, attitude, and/or behavior that results from the actual or imagined presence of a person or group of people" (Dambrun & Belletier, 2024). In Asch's study (1951), participants were invited to take part in what they believed to be a simple visual perception test. They had to publicly identify which of three lines drawn on a board corresponded to the length of a reference line. A naive participant was present, surrounded by accomplices who deliberately gave wrong answers. The results show that, on average, 33% of the naive subjects' responses aligned with the response given by the accomplices. A variation of the experiment shows that when naive participants are given the opportunity to give their responses privately rather than publicly, the rate of conformity to the responses given by the accomplices drops to 12.5%. The conformity effect can be explained by two processes highlighted by Deutsch and Gerard (1955): on the one hand, the process of informational influence, which consists of conforming to the opinions and behaviors of others in situations of uncertainty; on the other hand, the process of normative influence, which consists of conforming under normative pressure from the group. What about the effect of conformity on creativity? It has inhibitory effects: conformity induces pressure to conform to group norms and expectations, which can limit divergent thinking in creativity (Nemeth, 1986) and lead to homogenization of thought. Innovative ideas are then less likely to emerge ("groupthink effect," Janis, 1982). Nevertheless, conformity can also induce certain facilitating effects (Paulus & Nijstad, 2003) when it is valued. As a result, norms of creativity emerge and guide efforts in directions accepted by the group.

Diversity and Mix in Groups

A wealth of literature highlights the importance of diversity in promoting collaboration and creativity within teams. The interaction between diverse perspectives and collaborative efforts not only enriches the creative process, but also leads to innovative results that are often unattainable within homogeneous groups. Diversity or variety of perspectives stimulates creativity (However & van Knippenberg, 2021; Paulus et al., 2019a, 2019b). This diversity can relate to the field of expertise, age, training or education, or even the type of cognitive fixations that influence the ability to generate innovative ideas.

The meta-analysis by Hülsheger et al. (2009) shows that diversity related to job or professional functions (i.e., skills, expertise, knowledge of perspectives) is positively but moderately correlated with creativity (r =.155), while diversity related to personal characteristics (i.e., age, gender, origin) is negatively correlated ($r = -0.133$). The coefficient r here, and in the following presentations, corresponds to Pearson's correlation coefficient, which measures the strength and direction of the linear relationship between two variables (Cohen et al., 2013). Hundschell et al. (2021) showed in a systematic review of the literature (N = 119 studies) that diversity is an important asset in an organization. It thus helps to enhance creativity. To achieve this, organizations must foster inclusive environments that allow for the exchange of ideas from different perspectives. Wallrich et al. (2024), in a meta-analysis of 615 studies and 2,638 effect sizes, show a positive impact of diversity on creativity. Cognitive diversity (i.e., differences in knowledge, experience, and perspectives) improves the creative performance of teams, particularly for tasks that emphasize the divergence phase, as is the case with the creation of advertising slogans. Conversely, when tasks require convergent thinking, such as choosing a single solution, diversity can inhibit creativity. Leadership also plays an important role: research, such as that conducted by Roberson and Perry (2022), highlights the importance of relational leadership (i.e., establishing links with different teams, showing interest, and giving trust) in fostering a shared group identity and, beyond that, collaboration. For example, leaders can improve team dynamics by assigning projects to different members instead of relying on those whose work style and way of thinking are closest to their own (Downey et al., 2015; Jaiswal & Dyaram, 2019; Rock & Grant, 2016). The effects of diversity on creativity also vary depending on organizational and cultural contexts. For example, according to Salazar et al. (2017), the presence of members from different ethnic groups, combined with a strong shared team identity, promotes the generation of original ideas. A subtle balance between differences and similarities is therefore essential to maximize team creativity. The effects of diversity are also illustrated by an inverted U-shaped relationship, where too little or too much diversity can inhibit creative performance (e.g., Huang & Chen, 2010). Thus, diversity is an important lever for improving team creativity, but its effects depend on the context, the nature of the tasks, and the team's management style.

Diversity can also be expressed in terms of gender diversity. Gender diversity has a negative effect in the case of stereotypical tasks (Bell et al., 2011; Pearsall et al., 2008). Nevertheless, some research shows that women perform better in creativity (Bradley et al., 2020; Woolley et al., 2010). Peter et al. (2021) showed, in an electronic brainstorming exercise, where participants were asked to generate as many ideas as possible for recycling an object for new uses, groups composed solely of women were able to generate more ideas (fluency) and more original ideas than groups composed solely of men or so-called "solo" groups (i.e., with one member of one gender among a majority of the other gender). Furthermore, balanced mixed groups (with, e.g., two men and two women) performed comparably to groups composed solely of women.

Stereotypes

The activation of stereotypes can lead to performance that conforms to the expectations associated with those stereotypes. For example, evoking a belief about the lack of creativity of a group or community can impair the creative performance of its members compared to a situation where this stereotype is not activated. In this sense, Dumas and Dunbar (2016) compared the ability to propose innovative uses for everyday objects by asking their participants to place themselves in the shoes of a "strict librarian" or an "eccentric poet." These two professions—and the adjectives that were added—thus refer to personality stereotypes. Participants labeled with the "eccentric poet" stereotype (i.e., unique personality, unconventional, atypical lifestyle, etc.) showed significantly more originality and fluidity than those labeled with the "strict librarian" stereotype (i.e., strict and austere appearance, rigid and severe behavior). Thus, adopting a stereotype improves creative performance. The latter is considered malleable depending on the stereotypical context induced (see Tan et al., 2024; Wen et al., 2019; Zuo et al., 2019). In the field of digital technologies, avatars (digital representations of users) are becoming increasingly common in professional contexts (Ratan et al., 2022). Research highlights the positive effect of nonstereotypical avatars that resemble the user: De Rooij et al. (2017) show that an avatar similar to the user (wearing the same clothes) promotes creativity, unlike a stereotypical office worker avatar. In addition, Marinussen et al. indicate that identifying with an avatar with the characteristics of an individual considered creative can improve idea generation (see Bonnardel et al., Chap. 1 in this book).

Encouraging Creative Teamwork

In this potentially stimulating collective context, there are a number of creativity techniques, the best known of which is brainstorming, but there are also variants described in this section.

"Classic" Brainstorming

Brainstorming is a collaborative method developed by Alex Osborn (1953). It enables the generation of creative ideas by encouraging participants to express themselves freely and collaborate with each other. Brainstorming is based on a solid theoretical foundation (Paulus &

Kenworthy, 2019). Numerous studies have led to the development of theoretical models with many implications for collective ideation processes in organizational contexts (Paulus et al., 2018; Paulus & Nijstad, 2019). However, brainstorming situations are influenced by various psychosocial effects that are well known in the literature.

Group idea generation obviously relies on a process of sharing ideas with group members. Dugosh and Paulus (2005) and Paulus and Dzindolet (1993), for example, have shown that social comparison and exposure to the ideas of other group members promote the generation of additional ideas (see Fig. 3.4). However, brainstorming is also subject to certain effects that can stifle creativity. For example, social loafing encourages individuals to make less effort in a group than when they are alone. Fear or anxiety felt when being evaluated by others (Camacho & Paulus, 1995) is also a factor that inhibits creativity or causes "production blocking" (Diehl & Stroebe, 1987). Paulus and Brown (2007) proposed a cognitive-social-motivational model of group ideation. This model, shown below, presents the positive social influences that facilitate creativity and the negative ones that inhibit it.

Paulus and Kenworthy (2019) offer several tips for getting the most out of brainstorming, including: (1) induce a certain degree of social competition to strengthen participants' motivation; (2) formulate, before the

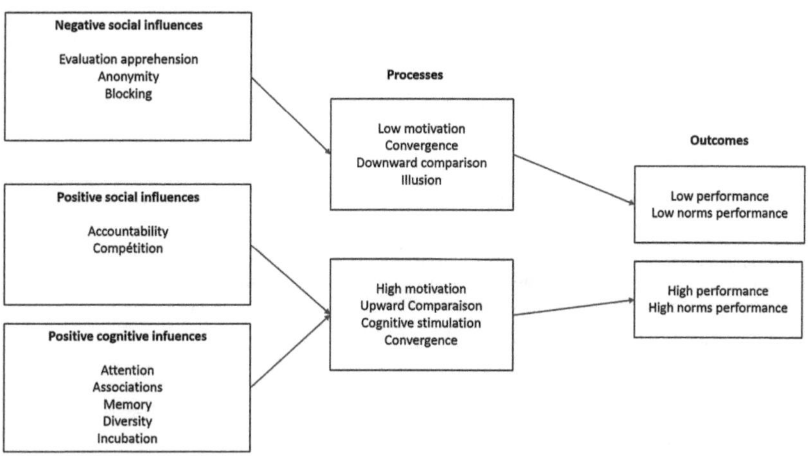

Fig. 3.4 The cognitive-social-motivational model of group creativity (based on Paulus & Kenworthy, 2019, p. 288; Bonetto & Girandola, 2023, p. 123)

session, a request that explicitly states, depending on the situation, the quantity and/or quality of ideas expected; (3) alternate between collective ideation sessions and individual sessions; and (4) integrate intellectual and cultural diversity into the group.

Paulus et al. (2019a, 2019b) also highlight specific principles for optimizing brainstorming. These principles find concrete application in the Divergent to Convergent Innovation Model (DCIM). This model analyzes the key factors for effectively linking the divergence and convergence phases of creativity in a collaborative setting (see Fig. 3.5). The divergent phase corresponds to the generation of ideas. It is influenced by the use of appropriate instructions, a relevant task structure (focus on categories, breaks), clear objectives, feedback, cognitive diversity, a positive mood, and openness to experience (Paulus & Kenworthy, 2019; Paulus et al., 2018). The convergent phase involves the selection and development of ideas. It is facilitated by appropriate instructions, cognitive diversity, conscientiousness, and a positive attitude toward group work (Fürst et al., 2016; George & Zhou, 2002; Lonergan et al., 2004; Putman & Paulus, 2009; Zhou & Su, 2010). The DCIM model emphasizes the importance of the links between these two phases: the quantity and quality of the ideas generated influence the quality of the final product. More specifically, the number of responses to shared ideas and the novelty of these responses significantly predict the novelty of the final creative product.

Several variants of brainstorming have been developed (e.g., Paulus et al., 2023; Paulus & Kenworthy, 2019): in groups or individually, in person, electronically (Baruah & Paulus, 2016; Maaravi et al., 2021), focused on idea generation or on generating constraints related to the

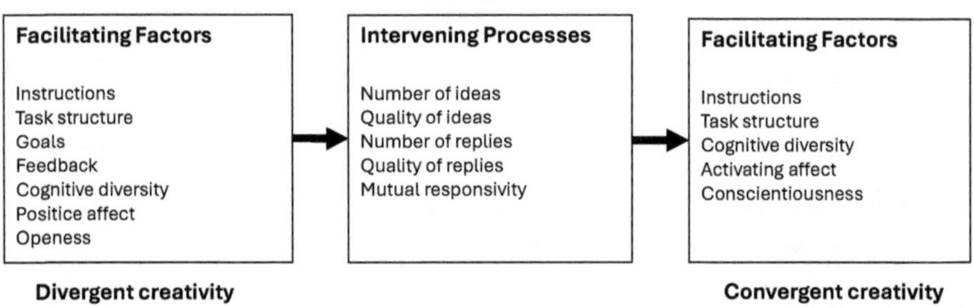

Fig. 3.5 A theoretical model of factors facilitating divergent and convergent collaborative creativity (DCIM), including intervening processes linking divergent and convergent creativity (based on Coursey et al., 2019a, 2019b, p. 3)

creative problem (e.g., Bonnardel & Didier, 2016, 2020). In the following section, we present different variants.

Variations on Brainstorming

- Brainstorming focused on ideas or focused on constraints related to the creative problem

A variant of brainstorming has been proposed to encourage participants, individually or collectively, to focus more on the constraints related to the creative problem before—or possibly during—idea generation (Bonnardel et al., 2013; Bonnardel & Didier, 2016, 2020). This variant of brainstorming, which focuses on constraints, is called CQHD because it encourages participants to consider and analyze the constraints (C) related to the creative problem prior to (or during) the generation of ideas. They are asked to generate a large quantity (Q) of constraints related to the problem or situation, then to prioritize them (H) and finally to multiply them (D). The aim is to encourage participants to explore and analyze the information and constraints that should be taken into account, so that these can guide them in solving the creative problem (Bonnardel, 2000; Bonnardel & Lellouche-Gounon, 2024). This objective is in line with research showing that low-performing teams focus on exploring solutions rather than analyzing the information needed to solve a problem (Sohrab et al., 2022, 2023). Thus, the aim is to avoid immediate fixation on solutions, which can be detrimental to the consideration and evaluation of available information and the constraints applicable to the situation.

Several studies (see, in particular, Bonnardel & Didier, 2016, 2020) have compared the effects of brainstorming focused on idea generation and brainstorming focused on constraint generation—also in comparison with a control situation (without brainstorming instructions). The results obtained showed different effects depending on the field of specialization or the type of training followed by the participants. For example, future designers benefit more from a constraint-focused variant, while future teachers benefit more from idea generation-focused brainstorming. Although such results also seem to depend on the complexity of the creative task, the choice of brainstorming variants should therefore be made according to the profile and background of the individuals required to demonstrate creativity (Bonnardel & Didier, 2020; Didier & Bonnardel, 2023).

- Electronic Brainstorming

In its simplest form, electronic brainstorming relies on the use of computer technology. It involves entering and disseminating ideas on a screen in real time to all members of a group. Individuals can consider the quality and/or quantity of ideas generated by others as a reference point for their own ideation process (Brown & Paulus 2002; Dugosh & Paulus 2005; Nijstad & Stroebe, 2006; Paulus & Dzindolet, 1993). Some research shows that the quality and quantity of ideas are highest in electronic brainstorming (Dennis et al., 2019; Guegan et al., 2021; Maaravi et al., 2021; Michinov, 2012; Michinov & Jeanson, 2021; Peter et al., 2021). This technique has several advantages: participants do not have to wait for their turn to speak, it allows ideas to be generated separately, and they can be shared at the same time while working together. Participants can enter their ideas anonymously, which helps to avoid evaluation apprehension (Aiken et al., 1995; Boddy, 2012; Goldenberg & Wiley, 2011). However, electronic brainstorming has some drawbacks: anonymity depersonalizes the discussion, which increases social loafing and dependence on the ideas and knowledge of others (Alnuaimi et al., 2010). Participants also tend to focus on generating their own ideas rather than evaluating and developing the ideas of others (Kay, 1995; Lindner, 1998). Furthermore, it is a slower channel of communication than direct speech, which can inhibit the generation of ideas (Dennis & Valacich, 1994; Dennis & Williams, 2003; Nunamaker et al., 1991; Williams & Karau, 1991).

- Intergroup Brainstorming

Paulus and Kenworthy (2020) and Kenworthy et al. (2024) present a variant of electronic brainstorming that explores the impact of exchanges between members of different groups. The study is based on 416 participants randomly divided into 106 groups. It is structured in four successive phases, beginning with a classic divergence phase in an individual setting, then progressing to group work. The first phase is individual brainstorming, where the goal is to generate as many ideas as possible individually online, without visibility between participants (see Fig. 3.6). The second phase consists of brainstorming where separate groups work on individual computers. The third phase allowed for a comparison of three modes of intergroup exchange: (a) the first mode allows free access and direct exchanges between all participants, (b) the second mode establishes that one member of each group acts as a mediator ("broker") and exchanges ideas with the other group (intergroup exchanges), (c) the third mode

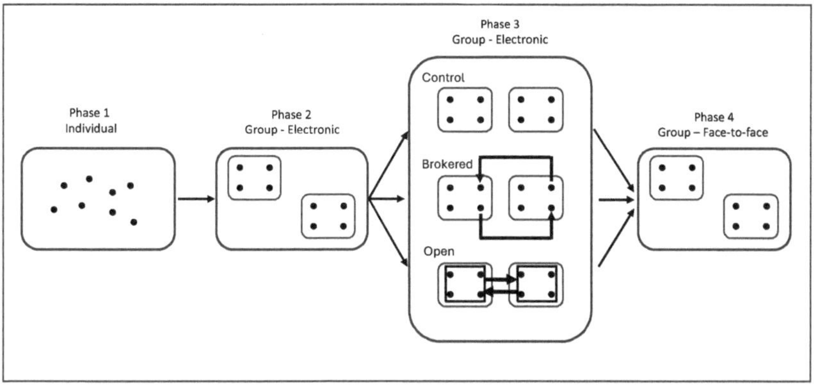

Fig. 3.6 Experimental design (based on Kenworthy et al., 2024, p. 1457)

corresponds to a control group without intergroup interactions. The fourth phase encourages individual creativity within the group and maximizes collaboration and diversity of ideas. Overall, the results show the superiority of the model with a mediator over free exchanges. To optimize these exchanges, it is necessary to organize structured interactions, train mediators, and favor long-term collaborative approaches tailored to the creative objectives of the groups.

Creative Team Performance: The Importance of Interaction and Context

Sohrab et al. (2022, 2023) and Zhao et al. (2024) formulated and successfully tested a number of recommendations aimed at improving team performance: (1) start meetings by reviewing all available information and sharing knowledge and information known to the team, (2) then review all possible solutions, (3) return to the available information, (4) seek confirmation from everyone, and (5) complete the task, ensuring that the decision is properly documented. These recommendations highlight the importance of information sharing and collective confirmation in promoting team performance. Other techniques, such as the de Bono's Hats technique, help structure exchanges within groups by encouraging participants to focus successively on different points of view when solving creative problems (Lellouche-Gounon et al., 2023; see Bonnardel et al., Chap. 1 in this book).

However, these structured exchange processes are weakened in a teleworking context due to the reduction in informal interactions, which are essential for creativity and collective innovation. Thus, some research mentions telework as an obstacle to team exchanges (see also Chap. 1). Does this mean that telework limits collective creativity? In the field of advanced engineering, where people are accustomed to working collectively using virtual design tools, Ferguson et al. (2024) have shown that teleworking among engineers involved in the development of complex aerospace systems during the COVID-19 pandemic is not conducive to the informal interactions that largely determine the richness and density of collective creativity and innovation. In this case, the loss of spontaneous discussions compromising social bonding and synergy within teams reduced the exchange of ideas and the flow of information, impacting productivity and collaboration. Rücker et al. (2024) analyzed the impact of the workplace on employee creativity according to task type among employees of a German company. The results show that for individual creative tasks (e.g., creating a presentation), employees are more creative when working remotely due to greater perceived psychological freedom. On the other hand, for creative team tasks (e.g., developing ideas for a new product), being in the same office is more conducive to creativity by reinforcing a sense of psychological security. These findings highlight the importance of adapting work environments to the specific needs of tasks in order to maximize employee creativity. This research echoes reflections on interpersonal distance between leaders and employees and between employees themselves (Meirovich & Goswami, 2021; Napier & Ferris, 1993).

Leadership and Creativity

Leadership Styles

Leadership is one of the factors that has the greatest impact on creative performance. Some research suggests complex interactions between different hierarchical levels (e.g., Dong & Tang, 2023; McKay & Kaufman, 2020; Mumford & Hunter, 2005; Reiter-Palmon, 2021). Mumford et al. (2023a, 2023b) conducted an analysis of different leadership styles on creative performance. They distinguish between leadership styles focused on human factors and those focused on organizational factors, and recommend the use of more objective creative performance indicators in leadership assessment.

There are four types of leadership focused on human factors, the main characteristics of which are as follows:

1. Servant leadership consists of serving others, with an emphasis on the development, well-being, and involvement of employees (i.e., listening, empathy, encouragement, altruism, shared responsibility, mutual respect, collaboration). Jaiswal and Dhar (2015) showed a correlation ($r = .42$) with creative performance (Liden et al., 2015). Wang et al. (2022) showed that servant leadership influences employee creativity ($N = 252$) via psychological safety (i.e., prioritizing employee needs, supporting their personal development, May et al., 2004) and their professional well-being (i.e., job satisfaction and fulfillment, Brunetto et al., 2011; Inam et al., 2021). Servant leadership therefore indirectly improves creativity through psychological safety and well-being (see Fig. 3.7).
2. Transformational leadership consists of encouraging, motivating, and inspiring subordinates to go beyond their personal interests and engage in creative activity at work. It involves valuing initiative. Gumusluoglu and Ilsev (2009) showed a relationship (ranging from $r = .19$ to $r = .45$) with creative performance. Qiao et al. (2025) show that transformational leadership stimulates employees to learn, question their assumptions, and adopt new and ultimately more creative perspectives on problems. This leadership style requires strong emotional regulation on the part of leaders in order to express

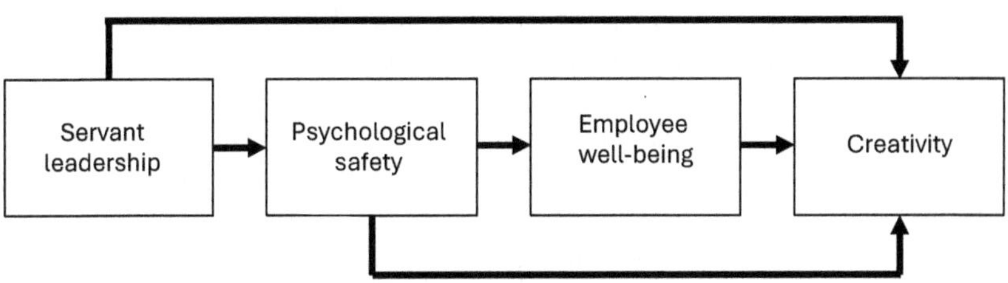

Fig. 3.7 Effects of servant leadership on creativity (based on Wang et al., 2022, p. 4)

enthusiasm and confidence. In the short term, this can be challenging, but in the long term, repeated regulation develops leaders' emotional self-efficacy. This increased confidence improves their own commitment to work and their ability to sustainably support their team (Bass & Riggio, 2006).

3. Empowering leadership. This involves trusting in the abilities of subordinates, encouraging decision-making, delegating responsibilities, supporting personal development, and encouraging innovation. Harris et al. (2014) demonstrated a relationship (r =.38) with creative performance (see Tse et al., 2023).
4. Authentic leadership. This involves establishing an egalitarian, authentic relationship with others, sharing values and beliefs in a climate of trust. Rego et al. (2012) demonstrated a relationship (r =.037) with creative performance.

In addition, there are two types of leadership focused on organizational factors:

5. Ambidextrous leadership. This involves managing the present and finding a balance between current operations (i.e., exploiting what we know how to do) and investing in new ideas. This refers to the ability of organizations to be both explorative and exploitative in order to remain competitive and successful in the long term (Zacher & Rosing, 2015). Ambidextrous managers encourage both the exploration of new ideas and the exploitation of current skills, which is a powerful combination for continued growth. Rosing and Zacher (2023) show a relationship (r =.45) with creative performance.
6. Entrepreneurial leadership. This involves identifying and exploiting opportunities and taking risks within the framework of leadership that is adaptable to the situation. Chen (2007) showed a relationship (r =.42) with creative performance.

Overall, leadership influences creativity with relationships ranging from r =.20 to r =.45, with an average of r =.30 (Mumford et al., 2023a, 2023b). The creative thinking of leaders thus has a direct impact on team innovation (Mumford & McIntosh, 2017; Thayer et al., 2018). The meta-analysis by Lee et al. (2020) examines 13 leadership styles and their differentiated impact on employee creativity (idea generation) and innovation (idea implementation), based on 266 studies. The results show that authentic leadership, empowering leadership, and entrepreneurial leadership are most closely linked to creativity. The study also highlights the importance of contextual conditions (e.g., industry, national culture, gender) in the development of creativity.

Innovative Work Behavior (IWB) is most often a determining factor in organizational development. Modiba et al. (2024) showed that among the factors influencing IWB, the quality of the relationship between the leader and their members (LMX theory or Leader-Member Exchange) plays a key role, although studies on this subject are sometimes contradictory. Research based on LMX theory has studied how this relationship impacts employee innovation through two elements: confidence in one's creative abilities (in line with motivation) and support for innovation in the team (in line with social relationship). The results, obtained from 171 employees and managers, show that leader-member exchanges promote innovation. However, this effect is mediated by support for innovation within the team, rather than confidence in one's creative abilities. Social dynamics within the team are therefore more important than individual motivations in encouraging innovation.

Furthermore, Cao et al. (2023) showed, in a sample of Chinese employees, that leaders' use of humor stimulates creativity. It promotes both incremental (incremental creativity) and major (radical creativity) improvements. These effects are more pronounced in public companies than in private companies due to better job protection and resources. Several studies have also shown the effect of leaders' humor on creativity in other populations (e.g., Pundt, 2024; Zhang et al., 2022).

While the quality of exchanges between leaders and team members directly influences creativity and innovation, the way individuals perceive leadership also plays an important role in organizational dynamics and creativity. Thus, a leadership style may not be imposed but may depend on how individuals or employees perceive it. Individuals use schemas, that is, a set of characteristics associated with a category, to simplify complex information. According to leader categorization theory, this same mechanism is used to perceive leaders (Lord & Shondrick, 2011). Individuals may also rely on prototypes or specific examples to form an idea of a leader (Zarate & Smith, 1990). Some research shows that it is important to understand how people represent an ideal leader prototype (Junker & Van Dick, 2014), that is, how a leader should be, rather than how a leader actually is. In the context of organizational creativity, these ideal prototypes have an effect on perception. For example, leaders who are perceived as open, innovative, and supportive of risk-taking are more likely to encourage (transformational leadership) the emergence of creative ideas within teams.

Leadership and Encouragement of Creativity

Leaders can also encourage creativity in their teams by implementing strategic planning (medium and long term) aligned with the organization's objectives, thereby creating an environment conducive to creative output (Mumford et al., 2008). This involves integrating creative thinking into training sessions for leaders in order to stimulate innovative approaches (McKay et al., 2024; Marcy & Mumford, 2007). Leaders must also know how to manage the cognitive diversity of their team members in order to foster creativity (Hunter et al., 2007; Medeiros et al., 2023). In addition, some studies aim to enable leaders to guide their interlocutors in order to promote "cognitive defixation" (Boudier et al., 2023) and thus lead individuals to orient their thinking to avoid a fixity or fixation effect. This effect occurs when individuals tend to remain fixated on unoriginal solutions based on easily accessible knowledge or examples provided by the experimenter, without managing to go beyond them to generate more creative ideas (Botella et al., 2023).

From Individuals to Organizational Creativity

Definitions and Levels of Organizational Creativity

In line with definitions of creativity that point out characteristics of creativity related, on the one hand, to novelty and, on the other hand, to adaptation, efficiency, or value (e.g., Bonnardel, 2002; Lubart et al., 2015; Runco & Jaeger, 2012), organizational creativity is defined as "the creation of a new product, service, idea, procedure, or process that is useful and valuable by individuals working together in a complex social system" (Woodman et al., 1993). According to Parmentier et al. (2018), it is also a "set of activities involving the creation, capture, selection, and implementation of new and useful ideas for the organization." Organizational creativity is the product of an interaction between individual creativity, collective creativity, and organizational characteristics (see Bonnardel et al., 2023; Carrier & Gélinas, 2016; Dechamp & Szostak, 2016).

Parmentier (2023) argues that there are different levels of creativity within a company: (1) the absence of creativity, when products and services are standardized; (2) creativity as a "trend," corresponding to a project temporarily assigned to certain people, with the use of creativity tools, occasional requests for input from creative individuals, or the development of coworking spaces, but without any real impact on the organization; (3) creativity as a "process," associated with the establishment of a dedicated department and a commitment to partnerships, agencies, and laboratories; (4) creativity as a "strategy," with the integration of creative individuals into the company and the promotion of creativity.

Person–Job Fit in Organizational Creativity

Organizational creativity relies in particular on the ability of individuals to produce new ideas. In this respect, person–job fit is a key lever, as it promotes individual creativity, which is the basis of collective creativity (Anderson et al., 2014). Some research has therefore looked at the effects on creativity of the fit between the characteristics of individuals (or staff) and those of their work environment (e.g., Seong & Choi, 2019). This research distinguishes the fit or alignment between several aspects:

(1) Job requirements (i.e., skills, knowledge, responsibilities) and, on the other hand, staff capabilities, which is referred to as "Demands Abilities-fit" (or DA-fit). A longitudinal study of 525 employees (Wang & Wang, 2018) showed that optimal DA-fit increases the likelihood of observing creative behavior at work by 43%, compared to 28% for average fit. Yulianti and Usman (2019) showed that optimal DA-fit also reinforces creative self-efficacy, contributing to better creative performance. Kim et al. (2019) showed that an increase in job demands that is not offset by staff capabilities reduces the perception of DA-fit and, beyond that, negatively affects creativity. Finally, Vleugels and Flatau-Harrison (2023), in a meta-analysis of 62 studies, observe a curvilinear relationship: optimal DA-fit (neither too low nor excessive) maximizes creative and innovative behavior.

Historically, Tierney and Farmer (2002) defined creative self-efficacy as an individual's belief in their ability to produce creative results in their professional environment. The results show that creative self-efficacy significantly predicts creative performance, beyond general work-related self-efficacy. Furthermore, Tierney and Farmer (2004) showed that high expectations of creativity among employees can stimulate their performance and creativity (i.e., the "Pygmalion effect"). Creative self-efficacy acts as a mediator between managers' expectations, their behaviors, and employees' perceptions of creative performance.

(2) Personal needs (i.e., psychological, emotional) and, on the other hand, the resources made available (Needs-Supplies fit or NS-fit). This concept is part of the Person–Job Fit theory (De Crom & Rothmann, 2018). Cable and DeRue (2002; in Boon & Biron, 2016) show that satisfying psychological needs (e.g., autonomy, feedback) through NS-fit increases creativity by 37%. Neubert et al. (2008) demonstrated an interaction between NS-fit (emotional support and resources) and DA-fit (technical skills) in stimulating creativity. In this sense, Wang et al. (2018) also identified managerial support as a key element of NS-fit, facilitating the sharing of knowledge necessary for creativity.

Finally, person–job fit appears to require a dynamic balance between technical or task-related skills (DA-fit) and psychological well-being in an environment that meets emotional needs (NS-fit), both of which are essential for creativity. This dual adjustment fosters an organizational climate conducive to the emergence and maintenance of creativity, which is essential for innovation and collective performance.

Constraints and Creativity

From a *cognitive point of view*, as evoked above (section "Variations on Brainstorming"), constraints are not considered only as limiting factors but, on the contrary, as essential guides in the creative process (Bonnardel, 2000, 2012; Bonnardel & Lellouche-Gounon, 2024; Tromp, 2022; Tromp & Sternberg, 2022a, 2022b; Tromp et al., 2024). The term "constraint" refers, here, to "any form of expression that identifies and encompasses the specifications and scope of the envisaged solution, and allows to delimit the perimeter of the individual's or the collective's choices in the creative process" (Bonnardel & Lellouche-Gounon, 2024, p. 82).

According to the A-CM model proposed by Bonnardel (2000, 2006/2012; see Fig. 3.8), constraints underpin and guide decision-making during creative activities, and a process of constraint management interacts with other processes, such as analogy-making (and ideas association), in order to lead individuals to reach creative ideas and solutions,

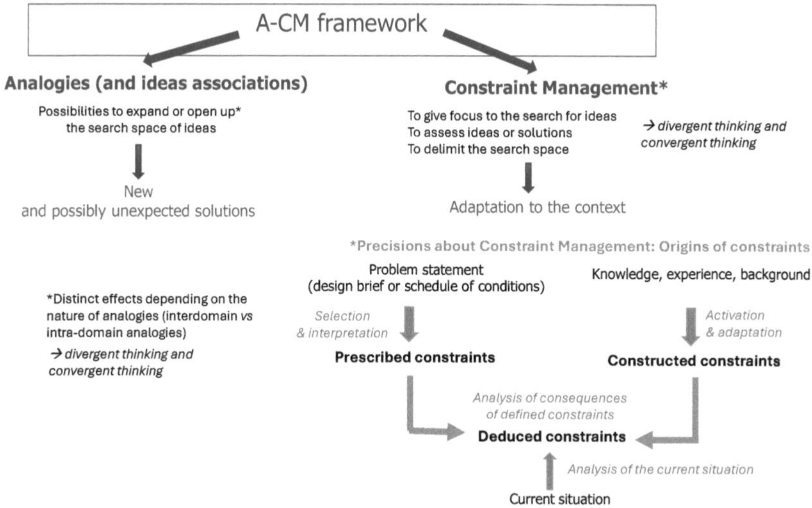

Fig. 3.8 The Analogies and Constraint Management model (based on Bonnardel, 2000; Bonnardel & Lellouche-Gounon, 2024)

which are both new—and, sometimes, unexpected—and adapted to the situation, and possibly even associated with value (Bonnardel, 2002). More precisely, in this theoretical framework (Bonnardel, 2000, 2006/2012; Bonnardel & Lellouche-Gounon, 2024), the constraints that intervene in creativity are not only external but also internal to each individual engaged in a creative activity, and these individuals can deal consciously or sometimes unconsciously with some constraints. *External constraints* include "prescribed" constraints that the individual defines himself or herself on the basis of an analysis—and selection—of the creative problem specifications (or schedule of conditions), but also on the basis of an analysis of technical or physical constraints present in the work environment (e.g., related to the available resources) as well as constraints relating to the organization of work (schedules, team composition). *Internal constraints* consist of so-called "constructed" constraints (e.g., Bonnardel, 2000, 2012; Bonnardel & Lellouche-Gounon, 2024; Bonnardel & Lubart, 2019) that individuals generate themselves, based on their previous experiences and background, possibly reflecting the influence of their childhood on the directions they choose to take in their creative activity, and/or depending on the values they wish to implement in the course of their creative activity (e.g., defending an environmental or inclusive perspective). Another kind of constraints, called "deduced," reflect the dynamic nature of creative activities in context (Bonnardel, ibid.) in the sense that they result from an analysis of the consequences of constraints already defined and an analysis of the current state of resolution of the creative problem.

In line with the A-CM model, Tromp (2022) proposed a model called IConIC (Integrated Constraints in Creativity) that proposes that creative outcomes emerge from the successful leveraging of different types of constraints. This author especially describes constraints that differ in terms of their flexibility (fixed, faux-fixed, or flexible) and which can give the individual a focus for the search based on exclusionary or focusing constraints. In addition, she proposes a distinction between constraints that apply to the problem search space and those that apply to the search time. Moreover, Tromp (2024) suggests that there is an inverted U-shaped relationship between creativity and constraints: too many constraints or too few constraints would not lead to creativity. This is in line with results of a meta-analysis focused on the role of constraints in creativity (Acar et al., 2019), which showed an inverted U-shaped relationship between various constraints and creativity.

From a psychosocial perspective, various constraints also apply to *organizational creativity* and are expressed in relation to individual–group interactions. Here, we discuss a few of the main ones: (1) social norms and pressure to conform, where dominant standards can limit creative thinking by encouraging individuals to adhere to ideas that are acceptable in their organizational environment, (2) the conscious or unconscious influence of stereotypes that can have a negative impact on creativity, (3) structural constraints with material limitations (i.e., time, budget, access to information) or organizational (rigid hierarchy) that can reduce freedom to explore and generate ideas, (4) self-censorship and internalization of conscious or unconscious limits (Cromwell et al., 2018, see Fig. 3.9), which appear as psychosocial barriers (i.e., lack of confidence, feelings of illegitimacy).

Some research shows that created or imposed limits can stimulate creativity by forcing people to think differently, outside the box, but this depends on the psychosocial context: if constraints are perceived as motivating challenges rather than threats, they promote creative adaptation. More generally, creativity most often emerges when these constraints are negotiated, circumvented, or transformed into levers (e.g., Bonnardel &

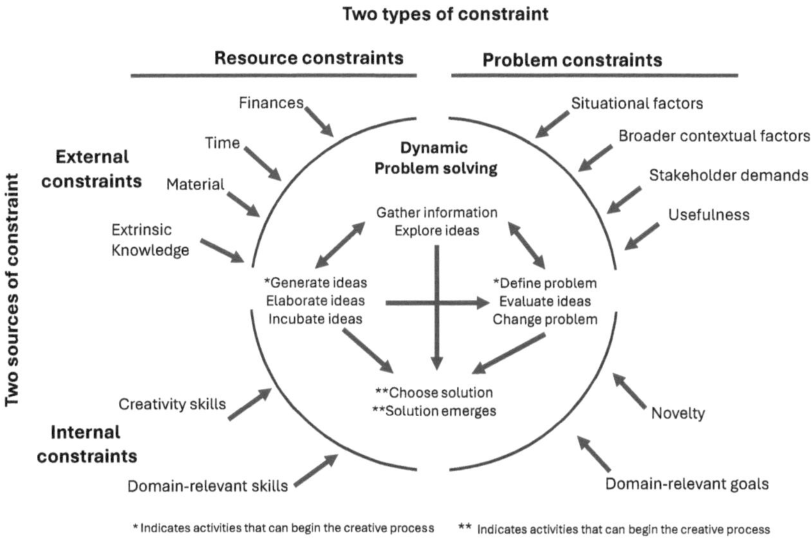

Fig. 3.9 Types and sources of constraints (based on Cromwell et al., 2018). Constraints are divided into two dimensions: horizontal (resources on the left, problems on the right) and vertical (internal at the bottom, external at the top). In the center, a dynamic problem-solving process unifies the "problem-first" and "idea-first" models. This model integrates these approaches while allowing for flexible iteration between them

Lellouche-Gounon, 2024). From this perspective, agency theory (Bandura, 2001) emphasizes the ability of individuals to act on their environment despite obstacles. Creative resistance (Scott, 1990) emerges when individuals perceive constraints as challenges. Finally, psychological reactance (Brehm, 1966; Brehm & Brehm, 1981) motivates individuals to reject perceived limitations in order to regain their freedom, thereby stimulating creativity. The theory of creativity under constraints (Stokes, 2006) shows that well-defined time and resource limits energize creativity rather than stifling it. Constraints also act as a motivational framework (Locke & Latham, 2002) that promotes focus and ingenuity. In short, psychosocial constraints do not kill creativity but redirect it. Medeiros et al. (2023) presented a sociocognitive model of constraints in which the following factors are equally important: a creative team climate, effective constraint management, psychological safety, trust, and reduced interpersonal conflict. In addition, Medeiros and Reiter-Palmon (2024) presented a model of diversity associated with constraints within a team: task-related diversity can influence constraint-related processes and, beyond that, the creative process.

Organizational Creativity at Apple (Lashinsky, 2012) or the Analysis of the Effects of Constraints on Creativity (Kocienda, 2018)
Apple has unique organizational creativity practices based on constraints. A leader, Steve Jobs, dictated the creative trajectory and oversaw everything from design details to the punctuation of marketing emails, concentrating creative power in the hands of executives with a big-picture view, but with teams working in silos, isolated and compartmentalized. For example, engineers working on the iPhone are unaware of parallel projects, and their access badges are disabled for areas outside their perimeter. The Industrial Design Studio, considered a creative sanctuary, is accessible only to an elite few. For Apple, design is the starting point for every product, and designers have creative veto power, with aesthetics and ergonomics taking priority over technology, forcing engineers to adapt. This approach has led to innovations such as the glass touch screen, which was developed in secret for years. Apple also cultivates what may seem like details, as even the packaging is designed to amaze. The designer responsible for iPod boxes tested hundreds of prototypes to find the perfect "feel" when opening them. This quest for perfection, high standards, and competitiveness has created internal pressure. Employees are evaluated on their ability to withstand stress and perform under pressure, such as during "crunch time" when everyone is working or mobilized for the launch of a new product. This method allows Apple to prioritize innovation over organizational constraints and maintain a constant flow of disruptive technologies. According to Lashinsky (2012), the key lies in the balance between creative freedom and rigorous control, where every detail is optimized to serve an overall vision. However, this method raises questions about managerial ethics. Apple legitimizes this pressure through the passionate commitment of its teams, transforming constraints into a rite of passage.

Parmentier (2023) shows the added value of owning equipment for organizational creativity: (1) cognitive equipment (i.e., individual and collective creative skills; common language; valuing creativity) and methods for developing creative abilities (C-K theory and DKCP method; Creative Problem-Solving; Design thinking, Rampa & Parmentier, 2023), (2) physical equipment that enables fertilization (i.e., creative atmosphere, collaboration and cross-functionality, support for discussing and selecting ideas, virtual spaces). Creative capacity is also an organizational construct that integrates strategic processes, routines, and resources.

Organizational Creativity at Pixar (Catmull, 2014)
Ed Catmull is President of Pixar Animation and Disney Animation

Ed Catmull's book shows the Pixar ecosystem. The organization relies on innovation through "brainstrust" or brainstorming. These are regular meetings between experienced peers: frank feedback is encouraged, and ideas that are implemented go through incubation phases with multiple versions of scenarios. "Notes Days" are days dedicated to sharing ideas among employees at all levels. These sessions aim to break down hierarchical silos and value diverse perspectives. Pixar benefits from a collaborative spatial environment that encourages chance encounters and spontaneous exchanges: shared cafeterias, modular rooms, and centralized restrooms. This approach has enabled Pixar to maintain an innovation rate 18% higher than the studio average (15 Oscars, 11 films including Toy Story grossing over $600 million) while avoiding the pitfalls of commercial pressure (see also Belsky, 2011).

Conflict and Conflict Management

Some task-related conflicts, such as disagreements between team members, can enhance creativity by encouraging them to engage in more in-depth discussions that allow for the exploration of new perspectives. Chung and Meneely (2012) showed that constructive disagreements can stimulate critical thinking and lead to innovative solutions, suggesting that a lack of conflict can stifle creativity (Barrett et al., 2021; Hirudayaraj & Matić, 2021). Diverse teams often experience higher levels of task-related conflict due to different perspectives and approaches (Aggrawal & Magana,

2024; Stel & van den Berg, 2022; Hundschell et al., 2021). These relational conflicts can also create a toxic environment, reducing trust and collaboration, which are essential to creative efforts (Ahmad et al., 2022). The challenge is to find a balance between the benefits of diversity and the risk of interpersonal friction that can undermine team cohesion and productivity. Benoliel and Somech (2014) showed that leaders can help team members appreciate their differences and transform potential conflicts into productive discussions. According to Shin and Eom (2014), transformational leadership enhances creative effectiveness, thereby facilitating a more collaborative approach (Han et al., 2017). Thus, while task-related conflicts are often beneficial to creative output, relational conflicts have a negative impact on team productivity. Effective leadership and team cohesion are factors that help manage these conflicts and foster a creative environment.

Creative Climate

Another important variable is the organizational creative climate, that is, the perception of a creative work environment. According to Paulus and Dzindolet (2008), a team climate that encourages support increases team performance and creativity, particularly by facilitating risk-taking (e.g., Reiter-Palmon, 2021). The climate affects several aspects of organizational life: team productivity, individual work and creativity through communication, coordination, learning, motivation, and problem solving (Tremblay & Simard, 2005). Hunter et al. (2007) proposed a model, based on a meta-analysis of forty-two studies, showing a positive relationship between organizational climate dimensions and creative performance. A creative climate is characterized by organizational support for new ideas, freedom to perform tasks, and recognition of creative work. The model also highlights the role of social factors such as trust, knowledge sharing, and intra-team communication. Moreover, this model identifies certain contextual characteristics such as sufficient resources, time allocated to creation, and rewards for creativity. All of these variables are predictors of creative performance, particularly in competitive environments. Several scales have been developed to measure climate (see da Silva Veiga & Cortez, 2022 for a comparison): (1) the KEYS scale “Assessing the climate for creativity” by Amabile et al. (1996) which proposes a 78-item scale measuring factors that stimulate creativity (i.e., encouragement of creativity, autonomy or freedom, resources, positive pressure), those that hinder it (i.e., negative pressure, organizational obstacles), and perceived productivity in organizations; (2) the "Creative Climate Questionnaire" (Ekvall, 1996), which comprises 50 questions to assess the organizational climate conducive to innovation and creativity (e.g., trust, challenge, motivation, support, freedom, pluralism, and debate); (3) the "Team Climate Inventory" (Anderson & West, 1998), which measures the climate at the team level with a focus on innovation and collaborative creativity; (4) the "Situational Outlook Questionnaire" (Isaksen et al., 2001), which is an evolution of Ekvall's Creative Climate Questionnaire (1996) across several dimensions (i.e., challenge, freedom, dynamism, trust, conflict, support), (5) the "Siegel Scale of Support for Innovation" (Siegel & Kaemmerer, 1978), which is based on 30 questions measuring leadership, ownership, diversity, and development (see Alencar & De Bruno-Faria, 1997); (6) the "Creative Environment Perceptions" (Mayfield & Mayfield, 2010); and (7) the "Indicators of conditions to create in the workplace" (Bruno-Faria et al., 2018).

Reflexivity

Reflexivity can be defined as the process by which team members analyze their experiences and practices together. It is an important lever for creativity in companies (Mol et al., 2015) and enables teams to adapt to organizational changes while maintaining their creative power and capacity for innovation (Köseoğlu et al., 2016). Reflexivity is stimulated by transformational leadership, which creates a climate conducive to the expression of innovative ideas (Dackert, 2016; Xu et al., 2024). It also helps to maintain a sense of psychological security, where everyone can freely share their ideas (Tang et al., 2024). Furthermore, it helps transform task-related conflicts into positive confrontation, thereby enhancing creativity (Chen et al., 2020). Finally, in virtual teams, reflexivity remains essential for maintaining cohesion and fostering remote innovation (Han et al., 2017).

Perception of Organizational Justice and Creativity

Some research shows the effect of organizational justice on creativity. Organizational justice refers to the perception of fairness and justice in the work environment. Traditionally, the literature (e.g., Colquitt et al., 2013;

Greenberg, 1987, 1993; Thibaut & Walker, 1975) distinguishes between three categories of justice:

(1) Distributive justice concerns the perception of resources, rewards, and sanctions. Employees assess whether salaries, promotions, and recognition are allocated fairly based on their contribution and performance.
(2) Procedural justice concerns the perception of fairness in decision-making processes and the rules used to distribute resources, rewards, and sanctions. Employees judge the transparency of processes, consistency in the application of rules, and the opportunity to express themselves or challenge decisions.
(3) Interactional justice concerns the perception of fairness in interpersonal interactions between employees and representatives of the organization. Employees judge the quality and fairness of interpersonal treatment by superiors and colleagues (respect, dignity, courtesy in interactions).

Sun et al. (2022), for example, show that climates of distributive, procedural, and interactional justice promote creativity through psychological safety and, above all, creative self-efficacy (Akram et al., 2022; De Clercq & Belausteguigoitia, 2020; De Clercq & Pereira, 2020, 2021; Khan et al., 2022; Khazanchi & Masterson, 2011). In an intercultural context, Sears et al. (2018) confirm that interactional justice intrinsically motivates and, beyond that, encourages creativity. Jaboob et al. (2023) highlight the positive impact of distributive justice on creativity and emphasize that certain leadership styles amplify this effect. Finally, Imamoglu et al. (2023) address the role of perceived organizational support in strengthening organizational identity and creativity, recalling that the perception of fairness and support from the employer remains a key factor in creative engagement. However, from an intercultural perspective, some findings qualify the impact of procedural justice (Jaboob et al., 2023), suggesting that the effects of perceived organizational justice on creativity may also depend on cultural contexts, leadership style, and work climate. Overall, research shows that a sense of organizational justice (vs. injustice) has a positive influence on creativity (Akram et al., 2022). Thus, the perception of organizational justice contributes to a climate of trust, psychological safety, and intrinsic motivation, which are essential conditions for the development of creativity.

Organizational Creative Capacity (OCC)

The concept of Organizational Creative Capacity (OCC) was developed by Parmentier et al. (2024), who proposed a scale for measuring organizational creative capacities. The study, conducted among 700 international respondents, identified five key dimensions or "organizational routines" that promote organizational creativity: (1) internal socialization of ideas related to the organization of continuous interactions between employees (Parmentier et al., 2023); (2) idea management, which includes the generation, selection, and evaluation of new creative ideas; (3) external organizational openness toward customers, suppliers, partners, and stakeholders, with the aim of creating new ideas (see Ruiz & Parmentier, 2023; Ruiz, 2024); (4) creative equipment aimed at stimulating creativity); (5) organizational agility referring to adaptation to change and monitoring of the environment. A version of this scale is available online: www.fabrique-crea.org/occ (Sheet et al., 2023). In addition, Szostak (2023a, 2023b) proposes a framework for analyzing the creative performance of organizations, including key indicators ("Creative ladder," creative KPIs) to assess its strategic role in transforming an organization and providing a sustainable competitive advantage (Brion & Ruiz, 2023).

Conclusion

Creativity in collective situations, particularly at the organizational level, is based on a complex interaction between social, cognitive, and organizational dimensions (Mumford et al., 2023a, 2023b). The aspects developed in this chapter—for example, social dynamics, leadership, and organizational creativity—illustrate how these aspects connect to foster an environment conducive to creativity and innovation. These different aspects have been structured according to different levels: individual, interpersonal or social interaction, group or collective, organizational, and methodological (see Table 3.1). Creativity thus develops in a structured ecosystem (Bonnardel et al., 2023).

This complexity also highlights several challenges to be addressed from a sociocognitive and psychosocial perspective: managing diversity and constraints, preventing cognitive and social blockages, and the need for effective leadership. According to Mumford et al. (2023a, 2023b), organizational creativity has been developing along several lines of research identified in the context of a new social psychology of creativity, including

Table 3.1 The main factors identified in this chapter that contribute to a socio-cognitive environment that stimulates creativity.

Level	*Factors promoting creativity*
Individual	– Intrinsic motivation (enjoyment, interest, curiosity, desire to excel) – Expertise and specialized skills – Creative skills (cognitive flexibility, originality, risk-taking, perseverance) – Cognitive style open to new perspectives – Creative self-efficacy
Interpersonal or social interaction	– Social consensus and recognition of ideas by the group, conformity – Positive influence of social comparison (mutual stimulation) – Limitation of inhibitory effects (social laziness, evaluation anxiety) – Avoidance of pressure to conform, which limit divergent thinking – Clearly established and valued norms of creativity
Group or collective	– Collaborative atmosphere and acceptance of failure – Motivational synchrony in teams – Adequate management of roles and skills within groups – Cognitive and cultural diversity of members – Balanced and cultural diversity of members (depending on the task) – Shared group identity – Constructive conflict management – Structured interaction techniques (brainstorming, Bono's hats)
Organizational	– Hierarchical and managerial support – Organizational culture that promotes innovation – Appropriate leadership (transformational, servant, empowering, authentic, ambidextrous, entrepreneurial) – Optimized person-job fit (DA-fit and NS-fit) – Cognitive and physical equipment (appropriate spaces, collaborative platforms) – Collective reflexivity (shared analysis of practices and experiences) – Strategic management of constraints (turning constraints into challenges) – Explicit encouragement of creativity by the organization (evaluation and promotion of new ideas)

(*continued*)

Table 3.1 (continued)

Level	*Factors promoting creativity*
Cross-cutting models, methodologies and techniques	– Amabile (1996), Paulus and Dzindolet (2008), Woodman et al. (1993) – Classic and electronic brainstorming (anonymity, real time, access to others' ideas) – Specific techniques (constraint-focused brainstorming, intergroup brainstorming) – Structuring creative activities into clear stages (exploration of constraints, idea generation, prioritization, selection) – Integration of constraints as a creative stimulus

social, motivational, and cognitive processes; leadership of creative teams; creative production methodologies; and individual differences in group and team creativity (Coursey et al., 2018). The Science of Teamwork, highlighted by Tannenbaum and Salas (2021) and his colleagues, seems to emphasize the importance of team processes in maximizing collective creativity. Their work shows that communication, coordination, and constructive conflict management are essential to the production of creative ideas. Integrating these elements makes it possible to better leverage cognitive diversity, strengthen team cohesion, and optimize creative performance, even in complex and uncertain contexts (McKay et al., 2021). To conclude this chapter, we suggest some concrete recommendations for optimizing creativity in collective situations, at the team and organizational levels, based on the nonexhaustive literature referenced in this chapter. In particular, we consider that it requires the following conditions:

(1) Creating a social climate conducive to creativity: promoting cognitive and cultural diversity; encouraging hybrid brainstorming practices (focused on ideas and/or constraints related to the creative problem, electronic or face-to-face brainstorming) to overcome production bottlenecks and leverage the benefits of digital tools; establishing a climate of psychological safety and organizational justice where employees feel free to propose ideas.
(2) Training and support for leaders and even middle managers: develop leadership skills, particularly "ambidextrous" leadership, to manage both the exploitation of existing ideas and the exploration

of new opportunities; encourage leaders to adopt servant or transformational leadership styles that stimulate motivation and creativity.

(3) Train and/or equip organizations to foster creativity: establish organizational routines, such as those identified in the OCC scale, including idea management, external openness, and organizational agility.

(4) Design physical spaces and digital environments that foster cross-functional collaboration, where ideas can be shared, discussed, and refined.

(5) Establish evaluation systems to value creativity and integrate measurable indicators.

(6) Leverage new technologies: invest in collaborative platforms and brainstorming tools (including, where appropriate, those based on interaction with generative AI) to maximize team productivity, explore virtual reality technologies to simulate creative contexts and strengthen personal and collective identity. These recommendations will provide organizations with a framework for transforming creativity into a strategic lever through well-structured collaboration.

References

Acar, O. A., Tarakci, M., & van Knippenberg, D. (2019). Creativity and innovation under constraints: A cross-disciplinary integrative review. *Journal of Management, 45*(*1*), 96–121.

Adarves-Yorno, I., Postmes, T., & Haslam, S. A. (2006). Creative and ingenious? The role of identity accessibility in the enhancement of creativity. *British Journal of Social Psychology, 45(3), 503–522.*

Aggrawal, S., & Magana, A. J. (2024). Teamwork conflict management training and conflict resolution practice via large language models. *Future Internet, 16*(5), 177.

Ahmad, A., Kee, D. M. H., & Daghriri, Y. Q. (2022). Conflict side of creativity: Role of supervisory support and team affective tone in facilitating creative idea validation. *Frontiers in Public Health.*

Aiken, M., Krosp, J., Shirani, A., & Martin, J. (1995). Electronic brainstorming in small and large groups. *Information & Management, 29(1), 35–40.*

Akram, Z., Ahmad, S., Akram, U., Asghar, M., & Jiang, T. (2022). Is abusive supervision always harmful toward creativity? Managing workplace stressors by promoting distributive and procedural justice. *International Journal of Conflict Management, 33*(3), 385–407.

Alencar, E. M. L. S. d., & Bruno-Faria, M. F. D. (1997). Characteristics of an organizational environment which stimulate and inhibit creativity. *The Journal of Creative Behavior, 31*(4), 271–281.

Alnuaimi, O. A., Robert, L. P., Jr., & Maruping, L. M. (2010). Team size, dispersion, and social loafing in technology-supported teams: A perspective on the theory of moral disengagement. *Journal of Management Information Systems, 27(1), 203–240.*

Amabile, T. M. (1983). *The social psychology of creativity.* Springer-Verlag.

Amabile, T. M. (1992). Social environments that kill creativity. In S. S. Gryskiewicz & A. D. Hills (Eds.), *Readings in innovation.* Center for Creative Leadership.

Amabile, T. M. (1993). Motivational synergy: Toward new conceptualizations of intrinsic and extrinsic motivation in the workplace. *Human Resource Management Review, 3*(3), 185–201.

Amabile, T. M. (1997). Motivating creativity in organizations: On doing what you love and loving what you do. *California Management Review, 40*(1), 39–58.

Amabile, T. M. (1998). How to kill creativity. *Harvard Business Review, 76*(5), 76–87.

Amabile, T. M. (2004). Motiver la créativité dans les organisations: faites ce que vous aimez et aimez ce que vous faites. In I. Getz (Ed.), *Créativité organisationnelle.* Vuibert.

Amabile, T. M., Barsade, S. G., Mueller, J. S., & Staw, B. M. (2005). Affect and creativity at work. *Administrative Science Quarterly, 50*(3), 367–403.

Amabile, T. M., & Conti, R. (1999). Changes in the work environment for creativity during downsizing. *Academy of Management Journal, 42*(6), 630–640.

Amabile, T. M., Conti, R., Coon, H., Lazenby, J., & Herron, M. (1996). Assessing the work environment for creativity. *Academy of Management Journal, 39*(5), 1154–1184.

Amabile, T. M., Hadley, C. N., & Kramer, S. J. (2002a). Creativity under the gun. *Harvard Business Review, 80(8), 52–61.*

Amabile, T. M., Hennessey, B. A., & Grossman, B. S. (1986). Social influences on creativity: The effects of contracted-for reward. *Journal of Personality and Social Psychology, 50*(1), 14–23.

Amabile, T. M., & Michael, G. P. (2016). The dynamic componential model of creativity and innovation in organizations: Making progress, making meaning. *Research in Organizational Behavior, 36*, 157–183.

Amabile, T. M., Mueller, J. S., Simpson, M. J., Constance, N.H., Kramer, S.J., & Fleming, L. (2002b). Time pressure and creativity in organizations: A longitudinal field study. *Harvard Business School Working Paper*, No. 02-073.

Amabile, T. M., Schatzel, E. A., Moneta, G. B., & Kramer, S. J. (2004). Leader behaviors and the work environment for creativity: Perceived leader support. *The Leadership Quaterly, 15*(1), 5–32.

Amabile, T. M., & Sensabaugh, S. J. (1992). High creativity versus low creativity: What makes the difference? In S. S. Gryskiewicz & A. D. Hills (Eds.), *Readings in innovation*. Center for Creative Leadership.

Anderson, N., Potočnik, K., & Zhou, J. (2014). Innovation and creativity in organizations: A state-of-the-science review, prospective commentary, and guiding framework. *Journal of Management, 40*(5), 1297–1333.

Anderson, N. R., & West, M. A. (1998). Measuring climate for work group innovation: Development and validation of the Team Climate Inventory. *Journal of Organizational Behavior, 19*(3), 235–258.

Asch, S. E. (1951). Effects of group pressure upon the modification and distortion of judgment. In H. Guetzkow (Ed.), *Groups, leadership and men*. Carnegie Press.

Asch, S. E. (1955). Opinions and social pressure. *Scientific American, 193*, 31–35.

Asch, S. E. (1956). Studies of independence and conformity: A minority of one against a unanimous majority. *Psychological Monographs: General and Applied, 70*(9), 1–70.

Bandura, A. (2001). Social cognitive theory: An agentic perspective. *Annual Review of Psychology, 52(1), 1–26.*

Baruah, J., & Paulus, P. B. (2016). The role of proximity in group creativity. *Journal of Experimental Social Psychology, 63, 41–52.*

Barrett, M. S., Creech, A., & Zhukov, K. (2021). Creative collaboration and collaborative creativity: A systematic literature review. *Frontiers in Psychology, 12*, 713445.

Bass, B. M., & Riggio, R. E. (2006). *Transformational leadership*. Psychology Press.

Bell, S. T., Villado, A. J., Lukasik, M. A., Belau, L., & Briggs, A. L. (2011). *Getting specific about demographic diversity variable and team.*

Belletier, C., & Camos, V. (2018). Does the experimenter presence affect working memory ? *Annals of the New York Academy of Sciences.* ISSN 0077-8923.

Belletier, C., Normand, A., & Huguet, P. (2019). Social-facilitation-and-impairment effects: From motivation to cognition and the social brain. *Current Directions in Psychological Science, 28*(3), 260–265.

Belsky, S. (2011). *Making ideas happen: Overcoming the obstacles between vision and reality.* Portfolio/Penguin.

Boddy, C. R. (2012). The nominal group technique: An empirical study of its group performance. *Australian Journal of Business and Management Research, 2(2), 52–64.*

Benoliel, P., & Somech, A. (2014). The role of leader boundary activities in enhancing interdisciplinary team effectiveness. *Small Group Research, 46*(1), 83–124.

Bonetto, E., & Girandola, F. (2023). Vers une (nouvelle) psychologie sociale de la créativité. In N. Bonnardel, F. Girandola, E. Bonetto, & T. Lubart (Eds.), *La créativité en situations* (pp. 116–131). Dunod.

Bonnardel, N. (2000). Towards understanding and supporting creativity in design: Analogies in a constrained cognitive environment. *Knowledge-Based Systems, 13*, 505–513. https://doi.org/10.1016/S0950-7051(00)00067-8

Bonnardel, N. (2002). Entrée: Créativité. In G. Tiberghien (Ed.), *Dictionnaire des Sciences Cognitives* (pp. 95–97). Armand Colin/VUEF.

Bonnardel, N. (2006/2012). *Créativité et Conception: Approches Cognitives et Ergonomiques.* De Boeck Supérieur.

Bonnardel, N., & Didier, J. (2016). Enhancing creativity in the educational design context: An exploration of the effects of design project-oriented methods on students' evocation processes and creative output. *Journal of Cognitive Education and Psychology, 15(1)*, 80–101. https://doi.org/10.1891/1945-8959.15.1.80

Bonnardel, N., & Didier, J. (2020). Brainstorming variants to favor creative design. *Applied Ergonomics, 83*, 102987. https://doi.org/10.1016/j.apergo.2019.102987

Bonnardel, N., Girandola, F., Bonetto, E., & Lubart, T. (Eds.). (2023). *La Créativité en Situations. Théories et Applications.* Dunod.

Bonnardel, N., & Lellouche-Gounon, A. (2024). Constraints in creativity: Their role and use to reach creative (design) solutions. In C. Tromp, R. J. Sternberg, & D. Ambrose (Eds.), *Constraint in Creativity* (pp. 77–99). Brill.

Bonnardel, N., & Lubart, T. (2019). La créativité: approches et méthodes en psychologie et en ergonomie. *RIMHE - Revue Interdisciplinaire, Management, Homme & Entreprise, 37*, 79–98. https://doi.org/10.3917/rimhe.037.0079

Bonnardel, N., Mazon, S., & Wojtczuk, A. (2013). Impact of project-oriented educational methods on creative design. In *Proceedings of the 31st European conference on cognitive ergonomics – ECCE 2013*, article no. 6. ACM Press.

Boon, C., & Biron, M. (2016). Temporal issues in person-organization fit, person-job fit and turnover: The role of leader-member exchange. *Human Relations, 69*(12), 2177–2200.

Botella, M., Pichot, N., Volle, E., Cassotti, M., Lubart, T., & Bonnardel, N. (2023). Créativité et cognition. In N. Bonnardel, F. Girandola, E. Bonetto, & T. Lubart (Eds.), *La créativité en Situations: Théories et Applications* (pp. 70–85). Dunod.

Boudier, J., Le Masson, P., & Weil, B. (2023). Accompagner la génération d'idées: expérimenter un leader défixateur en cas de fixations hétérogènes. In *Actes de la XXXIème conférence de l'AIMS* (p. hal-03701434).

Bouquet, C. A., Belletier, C., Monceau, S., Chausse, P., Croizet, J.-C., Huguet, P., & Ferrand, L. (2024). Join action with human and robotic co-actors: Self-other integration is immune to the perceived humanness of the interacting partner. *Quarterly Journal of Experimental Psychology, 77*(1), 70–89.

Bradley, B. H., Baur, J. E., & Banford, C. G. (2020). Creativity in teams: A review and future research agenda. *Journal of Management, 46*(5), 882–918.

Brehm, J. W. (1966). *A theory of psychological reactance. Academic Press.*

Brehm, J. W., & Brehm, S. S. (1981). *Psychological reactance: A theory of freedom and control. Academic Press.*

Brion, S., & Ruiz, E. (2023). L'articulation des capacités créatives et des capacités d'innovation. In N. Bonnardel, F. Girandola, E. Bonetto, & T. Lubart (Eds.), *La créativité en situations (pp. 216–228). Dunod.*

Brown, V. R., & Paulus, P. B. (2002). Making group brainstorming more effective: Recommendations from an associative memory perspective. *Current Directions in Psychological Science, 11*(6), 208–212.

Bruno-Faria, M. F., de Moura, G. L., & Veiga, H. M. S. (2018). Creativity in teams: A literature review. *Revista de Administração Contemporânea, 22(1), 84–104.*

Brunetto, Y., Farr-Wharton, R., & Shacklock, K. (2011). Supervisor-nurse relationships, teamwork, role ambiguity and well-being: Public versus private sector nurses. *Asia-Pacific Journal of Human Resources, 49, 143–164.*

Cao, Y., Zhou, K., Wang, Y., Hou, Y., & Miao, R. (2023). The influence of leader humor on employee creativity: From the perspective of employee voice. *Frontiers in Psychology, 14*, 1162790.

Camacho, L. M., & Paulus, P. B. (1995). The role of social anxiousness in group brainstorming. *Journal of Personality and Social Psychology, 68*(6), 1071–1080.

Carrier, C., & Gélinas, S. (2016). Le manager créatif: De la résolution de problèmes à l'innovation. Presses de l'Université du Québec.

Catmull, E. (2014). *Creativity, Inc. Overcoming the unseen forces that stand in the way of true inspiration*. Random House.

Chen, M. H. (2007). Entrepreneurial leadership and new ventures: A meta-analysis. *Journal of Business Venturing, 22*(5), 755–777.

Chen, C., Feng, J., Liu, X., & Yao, J. (2020). Leader humility, team job crafting and team creativity: The moderating role of leader–leader exchange. *Human Resource Management Journal, 31(1), 326–340.*

Chung, S. S. E., & Meneely, J. (2012). Profiling group dynamics within business and design student teams: Relationships among personality traits, problem–solving styles, and creative performance. *Journal of Interior Design, 37*(3), 23–46.

Clercq, D. D., & Belausteguigoitia, I. (2020). The links among interpersonal conflict, personal and contextual resources, and creative behaviour. *Canadian Journal of Administrative Sciences / Revue Canadienne Des Sciences De l'Administration, 38*(2), 135–149.

Coursey, L. E., Gertner, R. T., Williams, B. C., Kenworthy, J. B., Paulus, P. B., & Doboli, S. (2019a). Linking the divergent and convergent processes of collaborative creativity: The impact of expertise levels and elaboration processes. *Frontiers in Psychology, 12*(10), 699.

Coursey, L. E., Gertner, R. T., Williams, B. C., Kenworthy, J. B., Paulus, P. B., & Doboli, S. (2019b). Linking the divergent and convergent processes of collaborative creativity: The impact of expertise levels and elaboration processes. *Frontiers in Psychology, 10, 699.*

Cohen, J., Cohen, P., West, S. G., & Aiken, L. S. (2013). *Applied multiple regression/correlation analysis for the behavioral sciences. Routledge.*

Colquitt, J. A., Scott, B. A., Rodell, J. B., Long, D. M., Zapata, C. P., Conlon, D. E., & Wesson, M. J. (2013). Justice at the millennium, a decade later: A meta-analytic test of social exchange and affect-based perspectives. *Journal of Applied Psychology, 98*(2), 199–236.

Coursey, L. E., Paulus, P. B., Williams, B. C., & Kenworthy, J. B. (2018). The role of individual differences in group and team creativity. In R. Reiter-Palmon, V. L. Kennel, & J. C. Kaufman (Eds.), *Individual creativity in the workplace.* Academic Press.

Cromwell, J. (2020). The social psychology of creativity skills: A reconceptualization of the componential model. In R. Reiter-Palmon, C. M. Fisher, & J. S. Mueller (Eds.), *Creativity at work: a festschrift in honor of Teresa Amabile* (pp. 21–34). Palgrave Macmillan.

Cromwell, J. R., Amabile, T. M., & Harvey, J. F. (2018). An integrated model of dynamic problem solving within organizational constraints. In R. Reiter-Palmon, V. L. Kennel, & J. C. Kaufman (Eds.), *Individual creativity in the workplace.* Academic Press.

da Silva Veiga, H. M., & Cortez, P. A. (2022). Creativity climate: An analysis of measurement scales. In M. O. Macambira, H. Mendonça, & M. d. G. T. Paz (Eds.), *Assessing organizational behaviors.* Springer.

Dackert, I. (2016). Creativity in teams: The impact of team members' affective well-being and diversity. *Open Journal of Social Sciences, 04*(09), 19–12.

Dambrun, M., & Belletier, C. (2024). La conformité et l'obéissance. In L. Bègue-Shankland (Ed.), *Psychologie sociale: la science des interactions humaines* (pp. 323–352). DeBoeck.

De Clercq, D., & Pereira, R. (2020). Knowledge-sharing efforts and employee creative behavior: The invigorating roles of passion for work, time sufficiency and procedural justice. *Journal of Knowledge Management, 24*(5), 1131–1155..

De Clercq, D., & Pereira, R. (2021). Taking the pandemic by its horns: Using work-related task conflict to transform perceived pandemic threats into creativity. *The Journal of Applied Behavioral Science, 57(1), 104–124.*

De Crom, N., & Rothmann, S. (2018). Demands-abilities fit, work beliefs, meaningful work and engagement in nature-based jobs. *SA Journal of Industrial Psychology, 44*(1), 1–12.

Dechamp, G., & Szostak, B. (2016). Organisational creativity and the creative territory: The nature of influence and strategic challenges for organisations. *M@n@gement, 19*(2), 61–68.

De Rooij, A., van der Land, S., & van der Lugt, R. (2017). The effect of an example's modality on the originality of solution ideas. *Design Studies, 51*, 1–19.

Dennis, A. R., & Valacich, J. S. (1994). Group, sub-group, and nominal group idea generation in an electronic meeting environment. *Information Systems Research, 5(4), 365–388.*

Dennis, A. R., & Williams, M. L. (2003). Electronic brainstorming: Theory, research, and future directions. In P. B. Paulus & B. A. Nijstad (Eds.), Group creativity: Innovationthrough collaboration (pp. 160–178). Oxford University Press.

Dennis, A. R., Minas, R. K., & Wiiliams, M. L. (2019). Creativity in computer-mediated virtual groups. In P. B. Paulus & B. A. Nijstad (Eds.), *The Oxford handbook of group creativity and innovation* (pp. 253–269). Oxford University Press.

Deutsch, M., & Gerard, H. B. (1955). A study of normative and informational social influences upon individual judgment. *The Journal of Abnormal and Social Psychology, 51*(3), 629–636.

Didier, J., & Bonnardel, N. (2023). Créativité et didactique de la conception: Vers des méthodes d'idéation adaptées aux profils des apprenants. In N. Bonnardel, F. Girandola, E. Bonetto, & T. Lubart (Eds.), *La Créativité en Situations: Théories et Applications* (pp. 263–273). Dunod.

Diehl, M., & Stroebe, W. (1987). Productivity loss in brainstorming groups: Toward the solution of a riddle. *Journal of Personality and Social Psychology, 53*(3), 497–509.

Dong, Y., & Tang, M. (2023). Enable to innovate: A review of the influence of empowering leadership on follower creativity and innovation. In R. Reiter-Palmon & S. Hunter (Eds.), *Handbook of organizational creativity: Leadership, interventions, and macro levels issues (pp. 33–50). Academic Press.*

Downey, S. N., van der Werff, L., Thomas, K. M., & Plaut, V. C. (2015). The role of diversity practices and inclusion in promoting trust and employee engagement. *Journal of Applied Social Psychology, 45*(1), 35–44.

Dugosh, K. L., & Paulus, P. B. (2005). Cognitive and social comparison processes in brainstorming. *Journal of Experimental Social Psychology, 41*(3), 313–320.

Dumas, D., & Dunbar, K. N. (2016). The creative stereotype effect. *PLoS One, 11(2), e0142567.*

Ekvall, G. (1996). Organizational climate for creativity and innovation. *European Journal of Work and Organizational Psychology, 5*(1), 105–123.

Ferguson, S., van Velzen, E., & Olechowski, A. (2024). Team and communication impacts of remote work for complex aerospace system development. *Systems Engineering, 27*, 199–213.

Fillon, A. A., Girandola, F., Bonnardel, N., Kenworthy, J. B., & Souchet, L. (2024). Stage 2 Registered Report Personal factors and group creative outcomes: A correlational meta-analysis. *F1000Research, 13*, 904.

Ford, C. M. (1996). A theory of individual creative action in multiple social domains. *The Academy of Management Review, 21*(4), 1112–1142.

Fürst, G., Ghisletta, P., & Lubart, T. (2016). Toward an integrative model of creativity and personality: Theoretical suggestions and preliminary empirical testing. *Journal of Creative Behavior, 50*, 87–108.
George, J. M., & Zhou, J. (2002). Understanding when bad moods foster creativity and good ones don't: The role of context and clarity of feelings. *Journal of Applied Psychology, 87*, 687–697.
Goldenberg, J., & Wiley, J. (2011). Quality is not a substitute for quantity: The role of the number of examples in creativity. *Journal of Experimental Psychology: Learning, Memory, and Cognition, 37*(4), 1007–1014.
Greenberg, J. (1987). A taxonomy of organizational justice theories. *Academy of Management Review, 12(1), 9–22.*
Greenberg, J. (1993). The social side of fairness: Interpersonal and informational classes of organizational justice. In R. Cropanzano (Ed.), *Justice in the workplace: Approaching fairness in human resource management* (pp. 79–103). Lawrence Erlbaum Associates.
Guegan, J., Buisine, S., Nelson, J., & Vernier, F. (2021). Gamification and social comparison processes in electronic brainstorming. *Journal of Gaming & Virtual Worlds, 13*(3), 265–283.
Gumusluoglu, L., & Ilsev, A. (2009). Transformational leadership, creativity, and organizational innovation. *Journal of Business Research, 62*(4), 461–473.
Han, S. K., Chae, C., Macko, P., Park, W., & Beyerlein, M. (2017). How virtual team leaders cope with creativity challenges. *European Journal of Training and Development, 41(3), 261–276.*
Harris, J. I., Harvey, P., & Kacmar, K. M. (2014). The dark side of leadership: A multi-level model of the cascading effects of abusive supervision on subordinate creativity. *TheLeadership Quarterly, 25*(2), 299–312.
Haslam, S. A., Reicher, S. D., & Platow, M. J. (2015). The new psychology of leadership: Identity, influence and power. Psychology Press.
Hirudayaraj, M., & Matić, J. L. (2021). Leveraging human resource development practice to enhance organizational creativity: A multilevel conceptual model. *Human Resource Development Review, 20*(2), 172–206.
However, I. J., & van Knippenberg, D. (2021). How diversity promotes team creativity: Two bumpy roads to collective inspiration. In A. S. McKay, R. Reiter-Palmon, & J. C. Kaufman (Eds.), *Creative success in teams* (pp. 167–201). Academic Press.
Huang, Y. F., & Chen, Y. S. (2010). The impact of technological diversity and organizational slack on innovation. *Technovation, 30*(5–6), 420–428.
Hülsheger, U. R., Anderson, N., & Salgado, J. F. (2009). Team-level predictors of innovation at work: A comprehensive meta-analysis spanning three decades of research. *Journal of Applied Psychology, 94*(5), 1128–1145.
Hundschell, A. S., Razinskas, S., Backmann, J., & Hoegl, M. (2021). The effects of diversity on creativity: A literature review and synthesis. *Applied Psychology, 71*(4), 1598–1634.

Hunter, S. T., Bedell, K. E., & Mumford, M. D. (2007). Climate for creativity: A quantitative review. *Creativity Research Journal, 19*(1), 69–90.

Imamoglu, S. Z., Erat, S., & Turkcan, H. (2023). How organizational identity relates to knowledge sharing and creativity: Moderating effect of perceived organizational support. *Kybernetes, 52*(10), 4476–4449.

Inam, A., Ho, J. A., Zafar, H., Khan, U., Sheikh, A. A., & Najam, U. (2021). Fostering creativity and work engagement through perceived organizational support: The interactive role of stressors. *SAGE Open, 11*(3).

Isaksen, S. G., Lauer, K. J., Ekvall, G., & Britz, A. (2001). Perceptions of the best and worst climates for creativity: Preliminary validation of the Situational Outlook Questionnaire. *Creativity Research Journal, 13*(2), 171–184.

Jaboob, M., Salim Ba Awain, A. M., & Al-Ansi, A. M. (2023). Sustaining employees' creativity through organizational justice: The mediating role of leadership styles. *Social Sciences and Humanities Open, 8*(1), 100693.

Jaiswal, N. K., & Dhar, R. L. (2015). Transformational leadership, innovation climate, and performance in India: The meditating role of knowledge management. *Journal ofBusiness Research, 68*(7), 1439–1446.

Jaiswal, A., & Dyaram, L. (2019). Towards well-being: Role of diversity and nature of work. *Employee Relations, 41(1), 158–175.*

Janis, I. L. (1982). *Groupthink: Psychological studies of policy decisions and fiascos.* Houghton Mifflin.

Junker, N. M., & Van Dick, R. (2014). Implicit theories in organizational settings: A systematic review and research agenda of implicit leadership and followership theories. *The Leadership Quarterly, 25*(6), 1154–1173.

Kay, G. (1995). A positive theory of creative discovery. Journal of Business Venturing, 10(1), 1-16.

Kenworthy, J. B., Coursey, L. E., Dickson, J. J., Paulus, P. B., Rozich, B. C., & Marusich, L. R. (2024). The impact of intergroup idea exposure on group creative problem-solving. *Group Processes & Intergroup Relations, 27*(6), 1452–1473.

Khan, A. K., Bell, C., & Quratulain, S. (2022). Interpersonal justice and creativity: Testing the underlying cognitive mechanisms. *Management Research Review, 45*(12), 1627–1643.

Khazanchi, S., & Masterson, S. S. (2011). Who and what is fair matters: A multifoci social exchange model of creativity. *Journal of Organizational Behavior, 32*(1), 86–106.

Kim, M., Lee, C., & Lee, S. (2019). The effects of transformational leadership on creativity and innovation: The mediating role of psychological safety. *Journal of Leadership &Organizational Studies, 26*(4), 489–502.

Kocienda, K. (2018). *Creative selection: Inside apple's design process during the golden age of Steve Jobs. Picador.*

Köseoğlu, G., Shalley, C. E., & Herndon, B. (2016). Task context changes: Teams' maladaptive responses to unanticipated change. *Journal of Applied Social Psychology, 47*(4), 195–212.

Lashinsky, A. (2012). *Inside Apple: dans les coulisses de l'entreprise la plus secrète au monde*. Dunod.

Latane, B., Williams, K., & Harkins, S. (1979). Many hands make light the work: The causes and consequences of social loafing. *Journal of Personality and Social Psychology, 37*(6), 822–832.

Lee, A., Legood, A., Hughes, D. J., Tian, A., Newman, A., & Knight, C. (2020). Leadership, creativity and innovation: A meta-analytic review. *European Journal of Work and Organizational Psychology, 29*(1), 1–35.

Lellouche-Gounon, A., Gounon, A., & Bonnardel, N. (2023). Créativité des personnes âgées et de leurs aidants dans le cadre d'ateliers de créativité collaborative et participative. In N. Bonnardel, F. Girandola, E. Bonetto, & T. Lubart (Eds.), *La créativité en situations: Théories et applications* (pp. 313–323). Dunod.

Leonard-Barton, D., & Swap, W. C. (1999). *When sparks fly: Igniting creativity in groups*. Harvard Business Press.

Liden, R. C., Wayne, S. J., Meuser, J. D., Hu, J., Wu, J., & Liao, C. (2015). Servant leadership: Validation of a short form of the SL-28. *The Leadership Quarterly, 26*, 254–269.

Lindner, J. R. (1998). An examination of the nominal group technique in a business setting. Journal of Agricultural Education, 39(3), 54-61.

Locke, E. A., & Latham, G. P. (2002). Building a practically useful theory of goal setting and task motivation: A 35-year odyssey. American Psychologist, 57(9), 705-717.

Lonergan, D. C., Scott, G. M., & Mumford, M. D. (2004). Evaluative aspects of creative thought: Effects of appraisal and revision standards. *Creative Research Journal, 16*, 231–246.

Lord, R. G., & Shondrick, S. J. (2011). Leadership and knowledge: Symbolic, connectionist, and embodied perspectives. *The Leadership Quarterly, 22*(1), 2017–2222.

Lubart, T. I., Mouchiroud, C., Tordjman, S., & Zenasni, F. (2015). *Psychologie de la Créativité* (2nd ed.). Armand Colin.

Maaravi, Y., Heller, B., Shoham, Y., Mohar, S., & Deutsch, B. (2021). Ideation in the digital age: Literature review and integrative model for electronic brainstorming. *Review of Managerial Science, 15*, 1431–1464.

May, D. R., Gilson, R. L., & Harter, L. M. (2004). The psychological conditions of meaningfulness, safety and availability and the engagement of the human spirit at work. *Journal of Occupational and Organizational Psychology, 77*, 11–37.

Mayfield, M., & Mayfield, J. (2010). Developing a scale to measure the creative environment perceptions: A questionnaire for investigating garden variety creativity. *Creativity Research Journal, 22*(2), 162–169.

Marcy, R. T., & Mumford, M. D. (2007). Social innovation: A conceptual framework. In G. J. Puccio, M. Mance, & M. C. Murdock (Eds.), *Creative leadership: Skills that drive change* (pp. 203–231). Sage.

McKay, A. S., & Kaufman, J. C. (2020). The role of constraints in creativity: A comprehensive review. *Psychology of Aesthetics, Creativity, and the Arts, 14(1), 3–21.*

McKay, A. S., Reiter-Palmon, R., Coombes, S. M. T., & Coombs, J. E. (2024). A meta-analysis of creativity training in organizational settings. *Creativity and Innovation Management, 33(4), 587–602.*

McKay, A. S., Reiter-Palmon, R., & Kaufman, J. C. (2021). An interdisciplinary view on team creativity: Toward integration across fields. In A. S. McKay, R. Reiter-Palmon, & J. C. Kaufman (Eds.), *Creative success in teams* (pp. 412–428). Academic Press.

Medeiros, K. E., Damadzic, A., & Tromp, C. (2023). A sociocognitive perspective of constraints and creativity. In R. Reiter-Palmon & S. Hunter (Eds.), *Handbook of organizational creativity: Leadership, interventions, and macro levels issues* (pp. 161–170). Academic Press.

Medeiros, K. E., & Reiter-Palmon, R. (2024). Diversity and constraints: A multilevel model of team diversity and constraint identification and management in creative efforts. In C. Tromp, R. J. Sternberg, & D. Ambrose (Eds.), *Constraint in creativity* (pp. 128–144). Brill.

Meirovich, G., & Goswami, A. (2021). Psychosocial and tangible distance between a leader and a follower: The impact on dyadic relations. *Journal of Leadership Studies, 14*(4), 2–6.

Michinov, N. (2012). Is electronic brainstorming or brainwriting the best way to improve creative performance in groups? An overlooked comparison of two idea-generation techniques. *Journal of Applied Social Psychology, 42*, 222–243.

Michinov, N., & Jeanson, S. (2021). Creativity in scientific research: Multidisciplinarity fosters depth of ideas among scientists in electronic "brainwriting" groups. *Human Factors: The Journal of the Human Factors and Ergonomics Society, 65*(7), 1542–1553.

Modiba, R., Fischer, S. B., Treffers, T., & Welpe, I. M. (2024). Translating leader-member exchange to innovate work behaviour: The role of creative self-efficacy and team support for innovation. *Creativity and Innovation Management, 33*(4), 671–684.

Mol, E. D., Khapova, S. N., & Elfring, T. (2015). Entrepreneurial team cognition: A review. *International Journal of Management Reviews, 17*(2), 232–255.

Moscovici, S. (1984). *The age of the crowd: A historical treatise on mass psychology.* Cambridge University Press.

Mumford, M. D., & Hunter, S. T. (2005). Innovation in organizations: A multi-level perspective on creativity, knowledge, and leadership. *Research in Multi-Level Issues, 4, 9–73.*

Mumford, M. D., & McIntosh, T. (2017). Creative thinking in the real world: The role of creative problem-solving skills. In M. D. Mumford & S. E. Medeiros (Eds.), *Creativity and innovation in organizations (pp. 3–30). Routledge.*

Mumford, M. D., Medeiros, K. E., & Partlow, P. J. (2008). Creative thinking: Processes, strategies, and knowledge. *The Journal of Creative Behavior, 42(1), 30–47.*

Mumford, M. D., England, S., Newbold, T. R., & Fichtel, M. W. (2023a). The future of organizational creativity research: What's there, what's missing. In R. Reiter-Palmon & S. Hunter (Eds.), *Handbook of organizational creativity: Leadership, interventions, and macro levels issues (pp. 409–422). Academic Press.*

Mumford, M., et al. (2023b). Leader thinking, follower thinking leader impacts on follower creative performance. *Annual Review of Organizational Psychology and Organizational Behavior, 10*(1), 413–440.

Napier, B. J., & Ferris, G. R. (1993). Distance in organizations. *Human Resource Management Review, 3*(4), 321–357.

Nemeth, C. J. (1986). Differential contributions of majority and minority influence. *Psychological Review, 93*(1), 23–32.

Neubert, M. J., Kacmar, K. M., Carlson, D. S., Chonko, L. B., & Roberts, J. A. (2008). Regulatory focus as a mediator of the influence of initiating structure and servant leadership on employee behavior. *Journal of Applied Psychology, 93,* 1220–1233.

Nijstad, B. A., & Stroebe, W. (2006). How the group affects the mind: A cognitive model of idea generation in groups. *Personality and Social Psychology Review, 10*(2), 142–175.

Nunamaker, J. F., Dennis, A. R., Valacich, J. S., Vogel, D. R., & George, J. F. (1991). Electronic meeting systems to support group work. *Communications of the ACM, 34(7), 40–61.*

Osborn, A. F. (1953). *Applied imagination: Principles and procedures of creative problem-solving. Charles Scribner's Sons.*

Parmentier, G. (Ed.). (2023). *Les capacités créatives des organisations. Iste Editions.*

Parmentier, G., Le Loarne-Lemaire, S., & Rampa, R. (2023). La gestion et la socialisation des idées dans les organisations. In G. Parmentier (Ed.), *Les capacités créatives des organisations* (pp. 129–153). Éditions ISTE.

Parmentier, G., Sheet, Z., Jeannot, F., & Rampa, R. (2024). Development of a multidimensional scale to measure organizational creative capabilities. *Journal of Product Innovation Management, 41*(6), 1184–1209.

Parmentier, A., Le-Dain, M. A., & Claudel, A. (2018). Contraintes et créativité en conception: une revue de littérature. *Le travail humain, 81*(1), 1–27.

Paulus, P. B. (2000). Groups, teams, and creativity: The creative potential of idea-generating groups. *Applied Psychology. An International Review, 49(2), 237–251.*

Paulus, P. B. (2020). Reflection on the impact of Teresa Amabile on the development of the field of group creativity. In R. Reiter-Palmon, C. M. Fisher, & J. S. Mueller (Eds.), *Creativity at work: A festschrift in honor of Teresa Amabile* (pp. 109–120). Palgrave Macmillan.

Paulus, P. B., Baruah, J., & Kenworthy, J. B. (2018). Enhancing collaborative ideation in organizations. *Frontiers in Psychology, 9*, 2024.

Paulus, P. B., Baruah, J., & Kenworthy, J. (2023). Brainstorming: how to get the best ideas out the "group brain" for organizational creativity. In R. Reiter-Palmon & S. Hunter (Eds.), *Handbook of organizational creativity: Individual and group level influences* (pp. 373–389). Academic Press.

Paulus, P. B., & Brown, V. R. (2007). Toward more creative and innovative group idea generation: A cognitive-social-motivational perspective of brainstorming. *Social and Personality Psychology Compass, 1*(1), 248–265.

Paulus, P. B., Coursey, L. E., & Kenworthy, J. B. (2019b). Divergent and convergent collaborative creativity. In I. Lebuda & V. P. Glăveanu (Eds.), Handbook of social creativity research (pp. 245–262). Palgrave Macmillan.

Paulus, P. B., Coursey, L. E., & Kenworthy, J. B. (2021). Managing the key processes for team innovation. In A. S. McKay, R. Reiter-Palmon, & J. C. Kaufman (Eds.), *Creative success in teams* (pp. 30–61). Academic Press.

Paulus, P. B., & Dzindolet, M. T. (1993). Social influence processes in group brainstorming. *Journal of Personality and Social Psychology, 64*(4), 575–586.

Paulus, P. B., & Dzindolet, M. (2008). Social influence, creativity and innovation. *Social Influence, 3(4), 228–247.*

Paulus, P. B., & Kenworthy, J. (2019). Effective brainstorming. In P. B. Paulus & B. A. Nijstad (Eds.), *The Oxford handbook of group creativity and innovation* (pp. 287–305). Oxford University Press.

Paulus, P. B., & Kenworthy, J. B. (2020). Group and intergroup creativity. In L. Argote & J. M. Levine (Eds.), *The Oxford handbook of group and organizational learning* (pp. 389–409). Oxford University Press.

Paulus, P. B., & Nijstad, B. A. (Eds.). (2003). *Group creativity: Innovation through collaboration*. Oxford University Press.

Paulus, P. B., & Nijstad, B. A. (2019). *The Oxford handbook of group creativity and innovation*. Oxford University Press.

Paulus, P. B., van der Zee, K. I., & Kenworthy, J. (2019a). Diversity and group creativity. In P. B. Paulus & B. A. Nijstad (Eds.), *The Oxford handbook of group creativity and innovation* (pp. 33–49). Oxford University Press.

Pearsall, M. J., Ellis, A. P. J., & Evans, J. M. (2008). Unlocking the effects of gender faultlines on team creativity: Is activation the key? *Journal of Applied Psychology, 93(1), 225–234.*

Pellegrin, L., Chassery, L., Chaudet, H., Texier, G., & Bonnardel, N. (2021). Decision-making during nonroutine outbreak management: Toward an exploration of experts' creative decisions. *Applied Ergonomics, 90*, 103232. https://doi.org/10.1016/j.apergo.2020.103232

Peter, L., Michinov, N., Besançon, M., Michinov, E., Juhel, J., Brown, G., Jamet, E., Cherbonnier, A., & ProFan Consortium. (2021). Revisiting the effects of gender diversity in small groups on divergent thinking: A large-scale study using synchronous electronic brainstorming. *Frontiers in Psychology, 12*, 723235.

Pundt, A. (2024). Humor in leadership and employee creative and innovative behavior. *Current Opinion in Psychology, 55*, 101723.

Putman, V. L., & Paulus, P. B. (2009). Brainstorming, brainstorming rules and decision making. *Journal of Creative Behavior, 43*, 23–39.

Qiao, Y., Fan, P., Li, F., & Chen, T. (2025). *Learning and adaptation of transformational leaders: Linking transformation leadership to leader self-efficacy for emotional regulation and work engagement.*

Rampa, R., & Parmentier, G. (2023). Méthodes, outils et espaces pour révéler la créativité des collaborateurs. In G. Parmentier (Ed.), *Les capacités créatives des organisations* (pp. 41–56). Éditions ISTE.

Ratan, R., Beyea, D., Li, B. J., & Graciano, L. (2022). Avatar characteristics and social interaction in virtual worlds. *Journal of Computer-Mediated Communication, 27*(1), zmb001.

Rego, A., Sousa, F., Marques, C., & Cunha, M. P. (2012). Authentic leadership promoting employees' psychological capital and creativity. *Journal of Business Research, 65(3), 429–437.*

Reiter-Palmon, R. (2021). Leading for team creativity: Managing people and processes. In A. S. McKay, R. Reiter-Palmon, & J. C. Kaufman (Eds.), *Creative success in teams (pp. 86–127). Academic Press.*

Reiter-Palmon, R., Fisher, C. M., & Mueller, J. S. (Eds.). (2020). *Creativity at work: A festschrift in honor of Teresa Amabile.* Palgrave Macmillan.

Reiter-Palmon, R., & Hunter, S. (Eds.). (2023a). *Handbook of organizational creativity: Leadership, interventions, and macro levels issues.* Academic Press.

Reiter-Palmon, R., & Hunter, S. (Eds.). (2023b). *Handbook of organizational creativity: Individual and group levels influences.* Academic Press.

Roberson, Q., & Perry, J. L. (2022). Inclusive leadership in thought and action: A thematic analysis. *Group & Organization Management, 47*(4), 755–778.

Rock, D., & Grant, H. (2016). Why diverse teams are smarter. *Harvard Business Review, 4*, 2–5.

Rosing, K., & Zacher, H. (2023). Ambidextrous leadership: A review of theoretical developments and empirical evidence. In R. Reiter-Palmon & S. Hunter (Eds.), *Handbook of organizational creativity: Leadership, interventions, and macro levels issues* (pp. 51–71). Academic Press.

Rothouse, M. J. (2020). *A mindful approach to team creativity and collaboration in organizations.* Palgrave Macmillan.

Rücker, M., Pakos, O., Windschiegl, S., & Voigt, K. (2024). Working in the office or working from home: Where are employees most creative? *Creativity and Innovation Management, 33*(3), 518–529.

Ruiz, E. (Ed.). (2024). *The creative capabilities of open organizations.* Wiley.

Ruiz, E., & Parmentier, G. (2023). Construire l'ouverture pour développer les capacités créatives des organisations. In G. Parmentier (Ed.), *Les capacités créatives des organisations* (pp. 41–56). Éditions ISTE.

Runco, M. A., & Jaeger, G. J. (2012). The standard definition of creativity. *Creativity Research Journal, 24*(*1*), 92–96.

Salazar, M. R., Feitosa, J., & Salas, E. (2017). Diversity and team creativity: Exploring underlying mechanisms. *Group Dynamics: Theory, Research, and Practice, 21*(4), 187–206.

Scott, W. E., Jr. (1990). The effects of extrinsic rewards on intrinsic motivation: A critique. *Organizational Behavior and Human Performance, 45*(2), 217–238.

Sears, G. J., Shen, W., & Zhang, H. (2018). When and why are proactive employees more creative? Investigations of relational and motivational mechanisms and contextual contingencies in the east and west. *Journal of Applied Social Psychology, 48(11), 593–607.*

Seong, J. Y., & Choi, J. N. (2019). Is person-organization beneficial for employee creativity? Moderating roles of leader-member and team-member exchange quality. *Human Performance, 32*(3–4), 129–144.

Sheet, Z., Jeannot, F., & Parmentier, G. (2023). Mesurer les capacités créatives organisationnelles. In G. Parmentier (Ed.), *Les capacités créatives des organisations* (pp. 172–155). Éditions ISTE.

Shin, Y., & Eom, C. (2014). Team proactivity as a linking mechanism between team creative efficacy, transformational leadership, and risk-taking norms and team creative performance. *The Journal of Creative Behavior, 48*(2), 89–114.

Siegel, S. M., & Kaemmerer, W. F. (1978). Measuring the perceived support for innovation in organizations. *Journal of Applied Psychology, 63*(5), 553–562.

Sohrab, G., Uitdewilligen, S., & Waller, M. J. (2022). The temporal phase structure of team interaction under asymmetric information distribution: The solution fixation trap. *Journal of Organizational Behavior, 43*(5), 892–911.

Sohrab, S. G., Waller, M. J., & Uitdewilligen, S. (2023). Is your team caught in the solution fixation trap? *Harvard Business Review.*

Stel, M., & van den Berg, A. (2022). The Aha! moment: A psychophysiological investigation of insight. *Frontiers in Psychology, 13*, 843920.

Stokes, P. D. (2006). *Creativity from constraints: The psychology of breakthrough. Springer Publishing Company.*

Sun, L., Li, C., Pan, W. C., & Leung, A. S. (2022). Justice climate and employee creativity in the work uncertainty context: A cross-level investigation. *Asian Business & Management, 22(4), 1065–1093.*

Szostak, B. L. (2023a). Le management stratégique de la créativité organisationnelle. In N. Bonnardel, F. Girandola, E. Bonetto, & T. Lubart (Eds.), *La créativité en situations* (pp. 206–215). Dunod.

Szostak, B. L. (2023b). La performance créative des organisations. In G. Parmentier (Ed.), *Les capacités créatives des organisations* (pp. 173–190). Éditions ISTE.

Tan, N., Parashar, N., Ahmetoglu, G., & Harris, L. T. (2024). Investigating the relationship between stereotyping and creativity during marketing campaigns in marketeers and audiences. *Scientific Reports, 14*(1).

Tang, B., Han, Y., He, G., & Li, X. (2024). The chain mediating effect of shared leadership on team innovation. *Heliyon, 10*(3), e25282.8.

Tannenbaum, S., & Salas, E. (2021). *Teams that work.* Oxford University Press.

Thayer, A. L., Petruzzelli, A., & McClurg, C. (2018). Addressing the paradox of the team innovation process: A review and practical considerations. *American Psychologist, 73*(4), 363–375.

Thibaut, J., & Walker, L. (1975). *Procedural justice: A psychological analysis.* Lawrence Erlbaum Associates.

Tierney, P., & Farmer, S. M. (2002). Creative self-efficacy: Its potential antecedents and relationship to creative performance. *Academy of Management Journal, 45*(6), 1137–1148.

Tierney, P., & Farmer, S. M. (2004). The Pygmalion process and employee creativity. *Journal of Management, 30*(3), 413–432. https://doi.org/10.1016/j.jm.2002.12.001

Tremblay, M., & Simard, G. (2005). Le climat organisationnel et son impact sur la mobilisation des employés. *Revue de gestion des ressources humaines, 56*, 23–38.

Tromp, C. (2022). Integrated constraints in creativity: Foundations for a unifying model. *Review of General Psychology, 27*(*1*), 41–61. https://doi.org/10.1177/10892680211060027

Tromp, C. (2024). Creativity from constraint exploration and exploitation. *Psychological Reports, 117*(4), 1818–1843.

Tromp, C., & Sternberg, R. J. (2022a). Dynamic creativity: A person × task × situation interaction framework. *The Journal of Creative Behavior, 56*(4), 553–565.

Tromp, C., & Sternberg, R. J. (2022b). How constraints impact creativity: An interaction paradigm. *Psychology of Aesthetics, Creativity, and the Arts, 18(6), 948–959.*

Tromp, C., Sternberg, R. J., & Ambrose, D. (Eds.). (2024). *Constraints in creativity. Brill Publishers.*

Tse, H. H. M., Ashkanasy, N. M., & Kaur, S. (2023). Transformational leadership and creativity. In R. Reiter-Palmon & S. Hunter (Eds.), *Handbook of organizational creativity: Leadership, interventions, and macro levels issues* (pp. 7–19). Academic Press.

van Bezouw, M. J., van der Toorn, J., & Becker, J. C. (2021). Social creativity: Reviving a social identity approach to social stability. *European Journal of Social Psychology, 51*(2), 409–422.

Vleugels, W., & Flatau-Harrison, H. (2023). Curvilinear relationships in person-environment fit research. *Journal of Occupational and Organizational Psychology, 96*(3), 669–691.

Wallrich, L., Opara, V., Wesołowska, M., Barnoth, D., & Yousefi, S. (2024). The relationship between team diversity and team performance: Reconciling promise and reality through a comprehensive meta-analysis registered report. *Journal for Business and Psychology, 39*, 1303–1354.

Wang, W., Kang, S., & Choi, S. B. (2022). Servant leadership and creativity: A study of the sequential mediating roles of psychological safety and employee well-being. *Frontiers in Psychology, 12*, 807070.

Wang, Y., Liu, J., & Zhu, Y. (2018). Humble leadership, psychological safety; knowledge sharing, and follower creativity: A cross-level investigation. *Frontiers in Psychology, 9.*

Wang, X. H., & Wang, X. (2018). Transformational leadership and employees' creativity: The mediating role of work engagement. *Journal of Management & Organization, 24*(2), 273–289.

Wen, F., Zuo, B., Xie, Z., & Gao, J. (2019). Boosting creativity, but only for low creative connectivity: The moderating effect of priming stereotypically inconsistent information on creativity. *Frontiers in Psychology, 10.*

Williams, K. D., & Karau, S. J. (1991). Social loafing and social compensation: The effects of task importance and group performance. *Journal of Personality and SocialPsychology, 61*(4), 570–581.

Woodman, R. W., Sawyer, J. E., & Griffin, R. W. (1993). Toward a theory of organizational creativity. *Academy of Management Review, 18(2), 293–321.*

Woolley, A. W., Chabris, C. F., Pentland, A., Hashmi, N., & Malone, T. W. (2010). *Evidence for a collective intelligence factor in the performance of human groups. Science, 330*(6004), 686–688.

Xu, S., Yaacob, Z., & Cao, D. (2024). How transformational leadership promotes employee creativity within the context of environmental dynamism: A multi-level perspective. *International Journal of Organization Theory & Behavior, 27(2), 101–119.*

Yulianti, P., & Usman, I. (2019). Building employee creative performance: Through person-job fit, empowering leadership, and creative self-efficacy. *International Journal of Innovation, Creativity and Change, 9*(8), 167–177.

Zacher, H., & Rosing, K. (2015). Ambidextrous leadership and team innovation. *Leadership & Organization Development Journal, 36*(1), 54–68.

Zarate, M. A., & Smith, E. R. (1990). Person categorization and stereotyping. *Social Cognition, 8(2), 161–185.*

Zhang, Y., Yin, C., Akhtar, M. N., & Wang, Y. (2022). Humor at work that works: A multi-level examination of when and why leader humor promotes employee creativity. *Frontiers in Psychology, 13.*

Zhao, Y., Liu, T., Han, X., & Gui, H. (2024). Team decision-making interaction and performance: A behavioral process-based relationship study. *Small Group Research, 55*(6), 919–952.

Zhou, J., & Su, Y. (2010). A missing piece of the puzzle: The organizational context in cultural patterns of creativity. *Management and Organization Review, 6*, 391–413.

Zuo, B., Wen, F., Wang, M., & Wang, Y. (2019). The mediating role of cognitive flexibility in the influence of counter-stereotypes on creativity. *Frontiers in Psychology, 10.*

CHAPTER 4

Social Process Assessment for Recognizing Creativity: A Creative Behaviors and Processes Coding Scheme

Kyle Christenson, Payge Japp, Apurva Patel, Roni Reiter-Palmon, Matthew Cronin, Joseph A. Allen, Marissa Shuffler, and Joshua D. Summers

Introduction

In the last two decades, organizations, researchers, and policy makers have recognized that an important skill required by the workforce of the future is creativity (World Economic Forum, 2020, 2025). At the same time, much of the work performed will happen in teams as problems facing organizations are too complex to be solved by one individual (Kozlowski & Bell, 2008; Reiter-Palmon et al., 2017). The use of teams to solve complex problems suggests that it is critical to understand how teams solve problems

K. Christenson • M. Shuffler
Clemson University, Seneca, SC, USA
e-mail: kmc9@g.clemson.edu; mshuffl@clemson.edu

P. Japp • R. Reiter-Palmon (✉)
University of Nebraska at Omaha, Omaha, NE, USA
e-mail: pjapp@unomaha.edu; rreiter-palmon@unomaha.edu

N. Bonnardel (ed.), *Collaborations in Creativity*, The Seven C's of Creativity, https://doi.org/10.1007/978-3-032-07753-0_4

creatively and, more specifically, the cognition associated with team creative problem-solving (Reiter-Palmon & Paulus, 2019). In line with the previous chapters in this book (Chap. 1 by Bonnardel et al., Chap. 2 by Baker et al., and Chap. 3 by Girandola & Bonnardel), the purpose of this chapter is to develop a new coding scheme to allow researchers to code team meetings and discussions for creative cognitive processes. In addition to outlining the theoretical foundations and the development process of the coding scheme, we also provide detailed information, instructions, and concrete examples to aid in the use of the coding scheme for interested researchers.

Theoretical Underpinnings

Team Cognitive Processes Associated with Creativity

The increasing uncertainty facing organizations means that employees are confronted with more ill-defined and ambiguous problems than in the past. Ill-defined problems are characterized by multiple possible goals, multiple ways of solving the problem, and multiple possible and acceptable solutions. It is this ambiguity within ill-defined problems that allows for creative solutions to arise (Mumford et al., 1991).

Multiple frameworks exist to describe the cognitive processes involved in team-level creativity problem solving, but these frameworks often differ and sometimes contradict each other (e.g., Finke et al., 1992; Merrifield et al., 1962; Mumford et al., 1991; Osborn, 1953; Silverman, 1985; Sternberg, 1988). In response to this inconsistency, Reiter-Palmon and Illies (2004) conducted a review to identify common elements across these diverse models. Their review found that, despite the differences, a few key cognitive processes consistently appeared across all models:

A. Patel • J. D. Summers
University of Texas at Dallas, Dallas, TX, USA
e-mail: apurvapatel@floridapoly.edu; joshua.summers@utdallas.edu

M. Cronin
George Mason University, Fairfax, VA, USA
e-mail: mcronin@gmu.edu

J. A. Allen
University of Utah, Salt Lake City, UT, USA
e-mail: joseph.a.allen@utah.edu

problem construction, which involves defining and framing a problem to be solved (Reiter-Palmon & Robinson, 2009); idea generation, which consists of producing potential solutions or ideas to address the previously constructed problem; and idea evaluation/selection, which involves assessing generated solutions and choosing the most suitable one. These three processes are considered fundamental to the creative process in teams, each playing a distinct role.

Problem construction is the preliminary process of identifying and outlining the problem set to be solved. When engaging in problem construction, team members communicate to collectively break down and clarify relevant information used to define the parameters of a given problem, which significantly influences how the problem is approached and ultimately solved (Arreola & Reiter-Palmon, 2016; Goh et al., 2013; Reiter-Palmon et al., 1997; Rostan, 1997; Weingart et al., 2005). This process is particularly important to creative problem-solving because, as previously mentioned, ill-defined problems are often ambiguous, lacking clear boundaries and/or obvious solutions; variations in individual problem representations across team members can hinder collaborative efficiency and creative performance via process loss (Gish & Clausen, 2013; Leonardi, 2011; Weingart et al., 2010). Thus, effective problem construction can increase the sharedness of the team's shared understanding, which is critical for effective collaboration (Reiter-Palmon & Murugavel, 2018).

Idea generation is the process that involves brainstorming and developing potential solutions to the problem (Reiter-Palmon & Robinson, 2009). This process emphasizes the generation of a wide range of ideas, prompting divergent thinking and the exploration of different possibilities (Runco, 2010). Idea generation is critical, such that it creates a foundation of possible solutions that team members will eventually evaluate and select (Puccio et al., 2018).

While not distinguished in other models of creative problem-solving, it is more recently argued that idea elaboration, where ideas are expanded upon and refined in detail, is a distinct phase from idea generation (Cronin & Loewenstein, 2020; Perry-Smith & Mannucci, 2017). While idea generation produces a range of initial concepts, idea elaboration deepens these concepts and potentially transforms them into more comprehensive ideas. We argue this distinction is important because idea elaboration involves more critical thinking and may require making adjustments or modifications to the ideas generated. Elaboration brings new ideas to mind that can modify how people define the problem they are solving. Often the new ideas can be assimilated to fit the current assumptions. Sometimes the new ideas contradict existing assumptions, and thus require that the problem definition itself change to accommodate these new ideas.

Idea evaluation and selection is the stage where team members assess potential solutions that have previously been generated and elaborated upon, deciding whether an idea should be rejected, revised, or implemented (Mumford et al., 2002). In this phase, team members examine and decide upon criteria, such as originality, practicality, or quality, to be used to evaluate the attractiveness of each idea (Amabile, 1982; Reiter-Palmon & Robinson, 2009). Naturally, evaluation acts as a critical step in the creative problem-solving process because it directly impacts the solution that is selected for implementation (Mumford et al., 2002; Reiter-Palmon & Murugavel, 2018).

Previous research suggests that the quality of each process can dictate the quality of subsequent processes in a linear fashion, especially in structured settings (i.e., quality of problem construction → quality of idea generation → quality of idea evaluation → quality of idea selection) (Mumford et al., 1991; Reiter-Palmon et al., 2008). However, it is important to note that teams may not always follow these processes in a perfectly linear manner (Reiter-Palmon et al., 2008). Team members may revisit previous stages to refresh their understanding or to reevaluate previous decisions and concepts as new challenges emerge and new information about the problem is made relevant in later stages of the team's performance wave.

Considerations for Coding Creative Processes in Teams

Teams, particularly those that are heavily interdependent, tend to develop formalized knowledge structures, commonly referred to as shared mental models, as team members interact over time. These structures consist of shared understandings related to tasks, processes, and teamwork norms, and work to support team performance by reducing ambiguity/misunderstandings and improving effective communication within teams (Cannon-Bowers et al., 1993; Ward et al., 1999). With respect to a creative problem-solving context, task-related shared mental models include a common understanding of the problem that the team is addressing (often called a problem representation), the potential solutions proposed, and the criteria for evaluating those solutions. It is worth mentioning that the consistency of shared mental models across team members has also been linked to superior performance specifically in creative tasks (Mumford, 2011).

Due to the unique manifestation of this type of shared cognition in teams, Leone et al. (2023) argued that team-level problem construction, idea generation, and idea evaluation can and should be measured by behaviors that indicate the formation of these shared mental models. Specifically, Leone et al. (2023) successfully used problem-relevant verbal information sharing, or utterances, during team interactions to infer the building of team-level problem representations and subsequently were able to explore natural flows between problem construction, idea generation, and idea evaluation/selection throughout team interaction. Building on this work, we developed a novel coding scheme to capture these behaviors during team interactions, specifically focusing on problem construction, idea generation, idea elaboration, and idea evaluation and selection. Our scheme enables us to infer the development of team-level problem representations and to explore the natural transitions among these processes throughout team interactions. Specifically, our model adds on idea elaboration and incorporates the Stoplight Model of Creativity as part of the coding scheme.

The Stoplight Model of Creativity

The Stoplight Model of Creativity is a framework that aims to explain how individuals use their perspectives to tackle any given problem or issue (Cronin & Loewenstein, 2020). The model suggests that one can either use a currently held perspective, add to it, or change it based on the demands of the situation. The authors use the concept of "craft" to help specify what should and should not be considered "creativity." Craft is the knowledge and expertise contained within one's existing perspective; tools used to make sense of encountered issues and select appropriate responses. Creativity comes into play when one encounters a situation that requires them to venture beyond their established perspective to appropriately address. See Fig. 4.1 for the full model.

The "use" strategy involves applying existing knowledge, skills, and abilities (KSAOs) and problem representations to effectively address an issue and requires no interruption in task-oriented progress. Thus, "use" is represented by the "green light" in the stoplight metaphorical model.

The "add" strategy involves reflecting on one's perspective and incorporating additional, congruent concepts to enhance it. The aim is to bridge any misalignment between the situation and the existing perspective by acquiring complementary KSAOs. Further, alterations to the

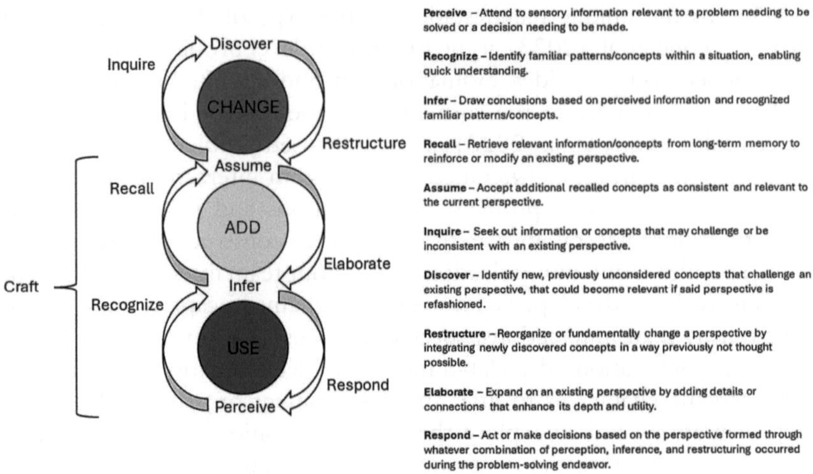

Fig. 4.1 The Stoplight Model of Creativity (adapted from Cronin & Loewenstein, 2020)

relevant problem representations are possible. This process of knowledge recall and elaboration may momentarily slow down or halt progress and is thus represented by the "yellow light." However, the changes to problem understanding are typically easy to incorporate. Both "use" and "add" are considered to be craft, as neither the perspective being used to address the problem nor the problem representation itself change in structure.

The "change" strategy entails questioning assumptions that were taken for granted in the problem representation for the situation at hand. This requires stopping task-oriented progress altogether and exploring new concepts that challenge the existing perspective(s) being used to address the problem. "Change" always starts with inquiry—the search for concepts that are inconsistent with one's given perspective and ends with discovery—the belief that the concepts one inquired about are appropriate to apply to the situation, which inspires the restructuring of one's perspective to integrate what was inconsistent to become consistent. This change in perspective necessitates a change in the associated problem representation. Thus, "change" is represented by the "red light" in the model, signifying the need to stop and engage in significant restructuring. By integrating new ideas, a previously inconceivable path towards problem resolution can emerge.

The authors of the model indicate that people are often most highly motivated to get a task done through a path of least resistance, which means that the assessment of fit between our perspectives and the issue at hand associated with the "change" strategy is often avoided unless deemed absolutely necessary. Thus, the "use" strategy is by far the most common to occur, followed by "add," and then to "change."

To help elucidate the model, the authors use an example of trying to solve a riddle. The specific riddle they use is as follows: "A man has married twenty women in this town, he has divorced none of them, and they are all still alive, yet he is not a polygamist. How is this possible?" The author's explanation has also been adapted into the following.

Use: Recognizing and Assembling a Familiar Story

When one first encounters the riddle, they generally begin by using their current perspective. One recognizes the individual concepts of "man," "married," "twenty," "women," and draws on our prior knowledge to perceive the situation. One infers that this is a story about a man in romantic relationships with twenty different women and responds by mentally constructing a familiar storyline: this man must be a polygamist.

This is all consistent with the default frame of "married" implying a spousal relationship. The concepts fit together—until they don't. The twist at the end—"yet he is not a polygamist"—contradicts the constructed perspective. There is a break in coherence: the story built from the current perspective no longer fits the facts.

Add: Attempting to Expand or Elaborate the Perspective

At this point, many people try to repair the mismatch by expanding their current perspective. They might recall alternative ideas or assume novel variations to make the story work. For example, one might suggest that the marriages were annulled quickly and didn't "count," or that "women" refers to multiple personalities in one woman. These moves elaborate the original perspective without fully abandoning it.

But even these attempted additions still do not quite hold together. One gets the sense that these explanations are strained or unsatisfying. Here, there is motivation to try to stay within the same basic interpretation, hoping that a clever twist within that original frame will resolve the riddle.

Change: Letting Go and Restructuring the Frame

Eventually, if no addition provides a satisfying solution, one must inquire—question whether one's entire frame might be flawed and look to discover new possibilities by considering that one of our base assumptions may be wrong. Could "married" mean something other than being a spouse? Could the man be involved in these weddings in another way?

Then comes the moment of insight: maybe he didn't *marry* the women in the romantic sense—maybe he *performed* the marriages. He's not the groom. He's the officiant. This insight creates a complete restructuring of the entire set of concepts. Suddenly, the story becomes simple and coherent: a priest has officiated twenty weddings; he's not a polygamist, he hasn't divorced anyone, and all the women are alive. Everything makes sense once the new perspective is adopted.

While this example does not specifically apply to a team problem-solving space, as the SPARC coding scheme does, it illustrates the individual-level cognitive work that is often present in those types of team interactions. Each team member brings a perspective shaped by prior experiences, expertise, and assumptions which they must use, elaborate, or revise in response to the problem at hand and the contributions of others. In collaborative settings, these perspective shifts happen through dialogue. One person might propose an idea using their current understanding (use), while another might build on it by introducing additional relevant information or connections (add). When progress stalls or contradictions emerge, the team may need to step back and reframe the problem altogether (change).

Comparable Instruments

The SPARC coding scheme can be viewed as the second iteration of a coding scheme developed by Leone et al. (2023), which was solely composed of codes related to the cognitive processes of creative problem-solving: problem construction, idea generation, and idea evaluation.

However, the structure of the SPARC coding scheme takes much more inspiration from the Act4Teams scheme (Kauffeld et al., 2018), another behavioral classification system used to make sense of team interactions, which itself takes from IPA (Bales, 1950), SYMLOG (Bales & Cohen, 1979), and KONFKOD (Fisch, 1994). The largest overlap between

SPARC and Act4Teams is the use of "sense units" as the base unit of analysis (Kauffeld et al., 2018).

Goals of the Instrument

Like its predecessors, the SPARC coding scheme is designed for use in group contexts where participants engage in problem solving efforts involving the exchange of different perspectives, the development of potential solutions, and agreement on paths forward. The tool holds three major advantages for the science of creativity. It allows researchers to:

1. Capture problem construction, idea generation, idea elaboration, and idea evaluation unobtrusively so as not to interrupt team interactions as they occur.
2. Determine the frequency with which teams engage in different creative problem-solving processes and to determine the amount of time teams spend in each "phase."
3. Identify meaningful sequences of behaviors (A → B → C), via sequential analysis, that are associated with higher levels of team creativity.
4. With regards to the Stoplight Model of Creativity, to distinguish "add" from "change." True insight comes from the "change," but often "add" is mislabeled as creativity.

Developing and Refining the Coding Scheme

Initial conversations held between Dr. Reiter-Palmon and Dr. Cronin, leading experts on creativity research, about the compatibility of various theories on cognitive processes related to creativity and the Stoplight Model of Creativity resulted in the initial matrix of codes seen in Table 4.1.

Both inductive and deductive methods were used for the process of refinement: Initial definitions of each code were created through conversations held between study authors, based largely in theories previously outlined in this chapter. A subteam of these researchers then used this initially proposed scheme to code a small subset of video recordings of teams engaging in a problem-solving activity.

The dataset used for coding scheme refinement was taken from an experimental study looking at the impact of team debriefs on the creativity of team problem solving efforts. In the experiment, teams of three met

Table 4.1 Exhaustive List of SPARC Codes and Abbreviations

Creative problem-solving process	*Stoplight model of creativity*		
	Use	*Add*	*Change*
Problem construction	PC-U	PC-A	PC-C
Idea generation	IG-U	IG-A	IG-C
Idea elaboration	IEl-U	IEl-A	IEl-C
Idea evaluation	IEv-U	IEv-A	IEv-C
Idea selection	IS	IS	IS

online using Microsoft Teams and were given prompts outlining set problem scenarios. Provided with 15 minutes for their activity, teams were asked to identify one overarching problem from the prompt and provide an appropriate solution. Responses were written down and submitted via online survey. Teams engaged in two problem-solving activities that were bisected by an intervention; either engaging in a team debrief or a control activity. The prompts provided for the two problem-solving activities can be found below.

Mark's Problem

Mark is a senior in college and captain of the swimming team. The swim team is at an out-of-town swim meet that is being held at another college. This is his first overnight trip as the captain of the team. The team consists of ten other swimmers and the coach. He found out that several team members are planning to sneak out of the dorm after curfew and meet with members of some of the other teams to go out to a bar. Mark's best friend seems to be the one organizing this night out, and he is generally a reliable person and does not get into trouble. Mark knows that the rules of the swim meet, as well as the rules of the team, prohibit leaving the dorm after curfew. The team could be disqualified from the meet if the officials find out. Mark is also worried about how well the team will perform in the meet tomorrow if they go out drinking tonight. If the coach finds out about this, the organizer of the outing will be suspended from the team for this meet, and he is one of the best swimmers on the team. Mark does not know what to do.

Space Hotel Problem

"Earth-Hilton" Staff Recruitment Information Brochure February 21st, 2027

Calling all those with an adventurous spirit. We are offering you the opportunity to work at the unique "Hilton Earth Orbit Hotel." We are looking for well-motivated, energetic people who want the adventure of a lifetime, working a long way from home in a zero-gravity environment. You need to be able to feel comfortable with spending long periods of time in confined spaces. Successful applicants will sign an initial contract for two years. On completion of your contract, you will be entitled to a free flight home. If, however, you decide to terminate your employment before the end of your contract you will be required to meet the costs of your homeward flight. You will be required to work six days a week.

Our guests pay $25,000 for a week's stay at the hotel so we aim to provide them with a very high standard of service. On your day off you are welcome to use the amazing facilities offered at the Hotel. We encourage you to take part in at least one of the gravisports on offer. You will be able to keep in contact with family and friends at home using the newly invented 3-D holographic communication system. It's almost as good as being with them in person.

Every two weeks you will be required to undergo an extensive medical examination. Our trained space physicians, who are experts in zero-gravity medicine and solar radiation exposure, monitor your fitness and health. We will monitor your bone density, muscle strength, immune system, and psychological health. Our hospital facilities are also equipped with an artificially intelligent space surgeon (AISS) which is programmed to operate in a zero-gravity environment. The AISS has proven to be 94% effective in the trials that have been run so far and it is becoming more proficient every day. All employees are required to attend the hotel gym for a minimum of one hour every waking cycle to complete a personalized routine of vibration and weight training based on the specially designed NASA health and fitness regime for people living in space. If you are female and planning on starting a family sometime in the future, we recommend that you have some of your eggs harvested and stored for this purpose. Some people have difficulty sleeping in a zero-gravity environment, even when strapped into their sleeping pod. We will provide sleeping tablets when required. Nano-cleaners remove human waste products and other pollutants from the air in the hotel.

Fresh food is provided from the biodomes in the hotel, but our first priority is to ensure that our guests are catered for to the standards expected in any of the Hilton hotels. You must be prepared to eat packaged food on occasion. We would ask that you eat only in the designated dining area. In lieu of paying for board, you will be required to pay taxes to WUSA (World United Space Agency) who have a role in waste removal and delivery of food, oxygen, and water to the Hotel. You will need to check with your own Government concerning further tax payments—some Governments require this, and others do not.

Since the transfer shuttle explosion during construction of the hotel, we have put in place an extensive disaster plan. Most problems will be repairable by our space-bots, but we have equipped the hotel with escape pods for both guests and employees though we doubt they will ever be required to be used. On initial acceptance of your application, you will be required to attend a four-week training session to initiate you into the rigors of working in a zero-gravity environment. Upon successful completion of this course, you will be welcomed as a new Hilton Earth Orbit team member.

Coding and Refinement Process

Video and audio recordings of these experimental team activities were initially cut and coded using the Act4Teams coding scheme, a popular scheme used to categorize and analyze information sharing between teammates during team meetings (see Kauffeld et al., 2018). The basic unit of measurement in these videos was the "sense unit," which represents one full idea expressed or utterance made by an individual (see details in *Unitizing* section). The process of "cutting" sense units involved time-stamping the beginning and end of each utterance and significant pause (operationalized a stretch of silence lasting at least 10 seconds) that occurred during each experimental team's problem-solving interaction and then labeling each sense unit using the Act4Teams scheme. All behavioral coding was conducted using Mangold InterACT qualitative analysis software, which was deemed appropriate given its ability to precisely time-stamp the beginning and ends of sense units, assign nominal codes to those sense units, and perform descriptive analysis of coded video data.

Coders tasked with using and refining the SPARC coding scheme began by establishing necessary Interrater Reliability, coding single videos independently and then meeting to review Kappa values between coding efforts. As many of the sense units were not directly related to creativity processes (see Table 4.2), and because this was a newly created coding scheme, researchers began by coding sense units in videos as either "null" or "code," indicating whether or not they believed a given sense unit could be appropriately labeled with a SPARC code. Null codes were then not coded further as they were not relevant to creativity processes.

Once it was determined adequate, interrater reliability had been established (Cohen's Kappa >.75) between all three coders and all "null coding" was complete, a second coding effort was then undertaken to label all "code" sense units from the first effort with the appropriate, specific SPARC code.

Any disagreement between coders about what a specific sense unit should be labeled, especially if it seemed to fall outside of initial code definitions, was discussed between the coders and the rest of the research team, leading to the definition refinement. Most of these conversations revolved around distinguishing between "add" and "change." Once consensus was reached, videos were re-coded using new rules until interrater reliability between the three independent coders again reached an appropriate level (Cohen's Kappa >.75).

The outcomes of the refinement efforts, including definitions and requirements, can be found in Tables 4.3, 4.4, 4.5, 4.6, and 4.7.

Tables 4.4, 4.5, 4.6, and 4.7 provide lists of general behaviors that would be categorized as "use," "add," and "change" with respect to

Table 4.2 Examples of "Null-Coded" Sense Units

Example "null" behaviors	*Reason for null coding*
"I'll read [the conflict scenario] first"	Procedural statement
"Do you want me to type [our answer]?"	Procedural statement
"What can a solution for that be?"	General discussion starter, not directed at a specific person
"Hmmm um let's see"	Filler words
"I don't know why they're calling it a...oh just kidding, nevermind"	Lacks subject and concrete contribution to team discussion
"Sorry...my audio is lagging so I can never tell when you're talking..."	Verbal behavior unrelated to problem discussion
Participant sighs	Verbal behavior unrelated to problem discussion
Pause in discussion (10+ seconds)	Pause in communication
Interruption	Break in communication

Table 4.3 Basic Descriptions of each SPARC code

SPARC	*Use*	*Add*	*Change*
Problem construction Specifying the nature of the task at hand via category	Restating what is already known about the problem in different ways	Building upon what is known about the problem in ways not previously discussed but that are consistent with the current understanding of the problem	Reframing or redefining the nature of the problem by introducing a new, potentially conflicting perspective that challenges the current understanding of the problem
Idea generation Stating actions to take or parts of the problem to change	Expressing known actions to alter known parts of the problem in expected ways	Expressing actions to take or parts of the situation to change that were not under consideration by all	Expressing actions to take or parts of the situation to change that do not previously make sense given the current prevailing perspective
Idea elaboration Building on an idea so that it is more descriptive or sophisticated while retaining its essence	The articulation of actions that could be taken and/or additional parts of an already proposed idea/solution that could be changed	The inclusion of other parts of an idea/solution or actions that could be taken that were not considered by all, but do not conflict with the current idea's structure	Altering the structure of an idea/solution for the use of an action in some way that was previously inconceivable given the current prevailing perspective
Idea evaluation Using current goals and norms to judge an idea	Statements about the value of an idea/solution relative to the team's goals	Statements that add parameters for evaluation of an idea/solution relative to the current situation	Statements that alter the criteria for evaluation in ways that seem incompatible with the current prevailing perspective
Idea selection Deciding which generated idea to move forward with/take action upon	Statements indicating agreement or disagreement with moving forward with a specific idea/solution	N/A	N/A
Null Any utterance/sense unit that does not meet the criteria to be labeled using one of the other categories	N/A	N/A	N/A

Table 4.4 Examples of problem construction codes

Use	*Add*	*Change*
Discussing/ summarizing/recapping the problem *E.g.: "So the problem we're trying to solve is [x]"*	**Framing the problem in various ways** *E.g.: "We could look at this as a communication issue, but it might also be a resource allocation problem"*	**Discussion indicating multiple competing ways to frame the problem** *E.g.: "I hadn't thought about the problem in that way before"*
Discussing confusion regarding the problem *E.g.: "I don't understand why this is a problem for him. Why is that an issue?"*	**Discussing task relevant information that other members missed in the problem** *E.g.: "Remember, in the problem statement, it said..." (when it is obvious another person missed an important piece of information about the problem)*	
Breaking the problem down into smaller pieces *E.g.: "Well, part of the problem is..."*	**Sharing past experiences relevant to the current problem** *E.g.: "I had this same problem when I was back in high school. It was really tough to deal with, but I chose to..."*	

Table 4.5 Examples of idea generation codes

Use	*Add*	*Change*
Discussion of possible solutions *E.g.: "A possible solution to this problem is..."*	**Suggesting solutions based on personal experiences** *E.g.: "If this were me, I would probably..."*	**Referring back to the problem and suggesting additional ideas** *E.g.: "Remember, we are trying to solve [x], so we might need to change the way we're approaching this idea"*
Asking team members about their perspective *E.g.: "What do you think we should do, [teammate]?"*	**Combining two suggestions together to form a new solution** *E.g.: "What if we took [teammate 1]'s idea and combined it with [teammate 2]'s approach?"*	**Proposing ideas that contradict or challenge assumptions held earlier in the discussion, but lead to a change in perspective** *E.g.: "We've been assuming we need to adapt [solution 1] to fit the current system, but what if we designed a completely new system instead?"*

Table 4.6 Examples of idea elaboration codes

Use	*Add*	*Change*
Asking another team member to expand on a suggested solution *E.g.: "That's an interesting idea, [teammate]. Can you explain a bit more about how it would work in practice?"*	**Suggesting additional details to an already proposed solution that introduce new elements or assumptions, thereby broadening what the solution includes** *E.g.: "Well, if the situation doesn't play out exactly like we think it will, we should add [a contingency plan]"*	**Suggesting a change in perspective to other team members** *E.g.: "Instead of looking at this problem from [x] angle, what might that [solution] look like from the perspective of long-term impact?"*
Suggesting additional details to an already proposed solution that add specificity or clarify how it would work, without expanding its original scope *E.g.: "We'd probably want to do [the idea] during the first week of being together, so everything stays on schedule"*		

problem construction, idea generation, idea elaboration, and idea evaluation. Specific examples of utterances relating to each categorical definition are also provided.

Practical Application

The following sections are included to provide additional, more specific guidance to researchers or practitioners who are interested in using the SPARC coding scheme.

General Coding Rules Specific to the Data Used for Scheme Refinement

This section outlines coding rules specific to the behaviors observed in the experiment that were used for refinement and validation. These may or may not apply to other experimental paradigms but are useful to

Table 4.7 Examples of idea evaluation codes

Use	*Add*	*Change*
Assessing the appropriateness of a solution for solving the problem *E.g.: "I don't think that [solution] would really solve the problem entirely"*	**Indicating additional criteria to be considered for idea evaluation** *E.g.: "We should also think about how easy it will be to implement [solution 1] with our current resources, not just how effective it might be"*	**Evaluating similarities and differences between suggested solutions** *E.g.: "That's similar/ different from the solution suggested earlier because..."*
Discussing/assessing the importance and prioritization of different criteria required to solve a problem well *E.g.: "I think we need to prioritize [criterion 1] over [criterion 2] for this problem since our main goal is to..."*	**Comparing ideas with respect to additional criteria** *E.g.: "[Solution 1] might save us more money upfront, but [solution 2] seems like it would require less maintenance in the long run. Which one aligns better with our overall goals?"* **with the understanding that maintenance is a newly introduced criterion**	**Using the same criterion but reaching different conclusions and reconciling** *E.g.: "I have not thought about it that way"*
Identifying pros and cons to suggested solutions *E.g.: "That's [not] a good idea, because..."*		

understand as they further illustrate initial conceptualization of many of the coding scheme's concepts.

Null

- Discussion of team procedure (e.g., how the team should interact and engage with the experimental task) was coded as null.

Problem Construction

- When a problem outside of the scope of the provided experimental prompt was brought up, it was coded as "Add." Researchers who had designed and conducted the experiment developed a list of all possibly identifiable problems specifically stated in the provided prompt, which was used to determine if a problem identified by a study participant was "within" or "outside the scope" of the prompt. All subsequent sense units referring to the "added" problem were then coded as "Use" because they were now within the scope of the team's shared mental model.

- General statements like "there are a lot of problems" were considered Null but saying "I don't see any problems" was coded as Problem Construction—Use.
- Solely affirmative words like "yeah" or "uh huh" are social processes and were thus coded as Null. However, if the affirmative words include an action (e.g. "yeah we should think about [the problem] that way") or reason (e.g. "yeah, it's important to remember that because...," we coded the sense unit as Problem Construction—Use.
- Encouraging general participation (e.g., "So what problems do we see?"), which was often observed at the beginning of each of the experimental tasks to start group conversation, was coded as Null. However, asking for someone's perspective about a specific problem was coded as Problem Construction—Use.
- Both questions clarifying what the provided prompt did or did not say as well as general responses to those questions were not coded as Problem Construction as there were no observable indications that participants viewed the situation being discussed as a problem to be solved. However, if the response to that type of question explicitly stated that it was a problem or included an evaluative statement, both the question and the response were coded as Problem Construction.

Idea Generation

- Like the rule for Problem Construction, when an idea/solution was initially brought up, it was coded as "Add" because the idea was being added from an individual's perspective to the team's shared mental model. However, all subsequent sense units referring to that idea/solution were coded as "Use" because it was now within the scope of the team's shared mental model.

Idea Selection

- Disagreement during Idea Selection was considered as Idea Evaluation, especially explaining why one disagrees with respect to the quality of an idea/solution.

- Idea Selection includes saying "yes" or "no" to a presented idea.
- Due to the nature of the experiment used for coding scheme refinement, participants would usually read aloud the answer they generated before submitting it. It was decided that both the reading aloud of the collective response by an individual and either affirmative or negative responses from teammates were all considered Idea Selection. This decision was made because these behaviors were deemed to be done with the intention of getting feedback from teammates regarding their approval of the response.

Examples of Coding

Below are two excerpts (see Tables 4.8 and 4.9) from transcripts pulled from the experiment used to refine and validate the SPARC coding scheme, along with the corresponding assigned codes.

Coding Time

If using a behavioral coding software, coding a single minute of recorded group interaction typically requires 5–10 minutes. However, this estimate can vary greatly depending on several factors including number of people speaking in the recording, the recording audio quality, the language or dialect being used by speakers, the complexity of the discussion topic, the sense unit density of conversation, and the level of experience of the coders.

Coder Training

Training coders typically takes between 80 and 100 hours, which should involve reviewing and discussing the code definitions with subject matter experts, practicing cutting and coding sense units with sample video/audio recordings, and cross-checking one's coding with the work of others on the same video/audio recording to ensure reliability. It is worth noting that the process of reaching rater agreement sufficient for the use of multiple coders may take substantially more time depending on the convergence process.

Table 4.8 Excerpt from Mark's problem

Transcript	*Speaker*	*Assigned code*	*Reason for code*
"Maybe Mark tells [the swim team], something they are forgetting, maybe something they're not thinking about very much, like the consequences of their actions, so maybe that can persuade them…eventually"	1	*IG-U*	*Introducing a new idea*
"Yes, and [Speaker 2], to go off of what you were saying earlier...so Mark is the swimming team captain..."	2	*Null*	*Not a complete thought*
"So, what he can try to do is talk to his best friend and explain exactly what's at stake…like the team getting disqualified or him being suspended…so he really understands the risk"	2	*IEl-U*	*Providing more details about what to say to the team, an idea previously expressed*
"And then, hopefully they'll...um stay in the dorm"	2	*Null*	*An ideal outcome of their idea, but not elaborative or evaluative. Expressed after a short pause from the last sentence*
"And if that doesn't work, Mark can always go to his coach and tell him what's going on"	2	*IG-U*	*Introduces the idea of getting the coach involved*
"Yeah, um, I'd agree, because, if he doesn't stop his best friend from going, they're all going to get in trouble and not perform well"	3	*IEv-U*	*Talking about the attractive benefits to the idea generated*
"Apparently his best friend is one of the best swimmers, so by telling… on him, he may save his swimming career"	3	*IEl-U*	*Elaborating on the idea to stop Mark's friends going out as it might have longer term consequences than just this swim season.*
"But, at the same time, they may lose the friendship for a brief period of time, or maybe forever, who knows…"	3	*IEv-A*	*Adding a potential consequence of telling the coach not previously discussed*
"I feel like the benefits outweigh the costs of him sneaking out"	3	*IEv-U*	*Prioritizing evaluatory criteria—stating some consequences are more important than others*

Table 4.9 Excerpt from Space Hotel problem

Transcript	*Speaker*	*Assigned code*	*Reason for code*
"So maybe, also with…the health, like with your checkup thing, also making sure that you are starting to try to go to the gym or go on walks and try to start to get active if you don't have a currently active lifestyle"	1	*IEl-A*	*Referring to a previously proposed idea about employees having more regular health checkups (not listed in this example) and suggesting more details such as other preventative measures to protect employee health*
"Yeah" (said passively)	2	*Null*	*Passive response (determined by vocal tone)*
"Do you have any potential solutions that you're thinking of, [speaker 3]?"	1	*IEl-U*	*Eliciting ideas from others (this was directed at speaker 3)*
"Um, I'm just taking notes right now, but...uhhh"	3	*Null*	*Procedural statement*
"I'm just thinking about the payment...one, uh, payment factor. Um..."	3	*PC-U*	*Bringing up another issue they found in the provided prompt*
"Maybe being a little more specific and saying if they're paying you in cash, like a cash settlement of some sort, or if… your [ability] to live there and not having to pay is your payment? Cause you're a worker"	1	*IG-U*	*Introducing potential ways of solving the issue of there being a lack of clarity regarding employee compensation*
"Mmmhmm, Mmhmm"	3	*Null*	*Passive response*
"Like…landlords sometimes have maintenance guys stay on the apartment site and live in an apartment for cheaper. It could be something like that. But maybe… just being more clear and specific"	1	*IEl-A*	*Suggesting additional details to an already suggested idea*

Technical Requirements

Coding using the SPARC scheme can be done without software assistance. However, we generally recommend using software to reduce the workload and streamline data processing and analysis. There are a number of publicly available software tools for doing this. For instance, as previously mentioned, InterACT software (Mangold, 2014) was used for SPARC development and refinement. In addition, there are other behavioral coding software options, such as the Communication Analysis Tool (Klonek et al., 2020) and BORIS (Friard & Gamba, 2016), which may also work depending on the researcher's goal. Still homegrown electronic coding systems and manual coding are perfectly fine ways of proceeding with analysis. For instance, artificial intelligence using large language models may be another tool that could help in this process, though we do not have experience using these types of tools in this manner.

Unitizing

There are generally two options for unitizing video and audio data for the purpose of interactional analysis: interval coding and event coding. Interval coding involves breaking a video/audio recording into predetermined, equally long segments (e.g., 10 second segments), and then assigning a code to each of those segments based on the most predominantly featured behavior displayed. While interval coding is a quick process, especially with the help of interactional analysis software, and may be sufficient for some research questions, one can lose some nuance in their data, as more than one meaningful behavior can occur during an allotted segment.

Conversely, event coding, which our team used for SPARC coding scheme development and refinement, involves manually assigning a beginning and ending time stamp to any and all behaviors that occur during a given recording. Units created with event coding are often referred to as "thought units" or "sense units," and represent the smallest, and in this case verbalized, behavior that meaningfully contributes to the group's conversation, sometimes even being as short as one word (see Kauffeld et al., 2018 for more details). It is recognized that in the context of creative problem-solving, "meaningful contributions" can be difficult to determine and may be context dependent. Thus, it is anticipated that coding rules may need to be slightly adjusted to better fit the nature of the problem-solving environment/context for which a researcher is using the SPARC codebook (see General Coding Rules section for contextually based rules used during codebook refinement).

Data Aggregation and Analysis

Each unit available for analysis specifies the exact beginning and end of the sense unit (hours, minutes, seconds, and frames per second), the speaker, and the corresponding behavioral code. Thus, initial analyses using SPARC can generate summative results including:

1. **Raw counts of behaviors**—For example, Team 1 engaged in PC-U 17 times but never engaged in a "change" behavior throughout their interaction.
2. **Time spent on specific behaviors**—This can be analyzed in total time per behavior or as relative percentages of the total interaction. For instance, Team 1 engaged in 23 minutes of problem construction, which accounted for 16% of their meeting.
3. **Change in count or frequency behavior over time**—For example, in the first quarter of the meeting, Team 1 spent 40% of their time on problem construction and 12% on idea generation. In the last quarter of their meeting, the team spent only 4% of their time on idea generation and 67% on idea evaluation.

Further, it is possible to analyze lags (sets of two behaviors) and sequences (sets of three or more behaviors) of coded behaviors. Lag analysis involves examining the likelihood of one behavior occurring directly after another in a given interaction. For example, a lag analysis might reveal that "PC-U" is consistently followed by "IG-A" within Team 1's interaction, suggesting a meaningful pattern in their problem-solving approach. Sequence analysis allows researchers to go one step further and uncover more complex patterns of interaction by identifying recurring chains of behaviors. For instance, a sequence analysis could show that the chain of "IEl-U → IEv-A → IS" appears more often in certain kinds of teams than others.

Concusion

In this chapter we have presented a new coding scheme for team interactions, one that focuses on creative problem-solving processes. We then discussed the details of procedures to follow when using the coding scheme and provided detailed examples of interactions and how these would be coded. This coding scheme can be used to identify creative cognitive processes in team interactions and facilitate our understanding of team creativity.

References

Amabile, T. M. (1982). Social psychology of creativity: A consensual assessment technique. *Journal of Personality and Social Psychology, 43*(5), 997–1013. https://doi.org/10.1037/0022-3514.43.5.997

Arreola, N. J., & Reiter-Palmon, R. (2016). The effect of problem construction creativity on solution creativity across multiple everyday problems. *Psychology of Aesthetics, Creativity, and the Arts, 10*(3), 287–295. https://doi.org/10.1037/a0040389

Bales, R. F. (1950). *Interaction process analysis: A method for the study of small groups.* Addison-Wesley.

Bales, R. F., & Cohen, S. P. (1979). *SYMLOG: A system for the multiple level observation of groups.* Free Press.

Cannon-Bowers, J. A., Salas, E., & Converse, S. (1993). Shared mental models in expert team decision making. *Individual and Group Decision Making: Current Issues, 221*, 221–246.

Cronin, M. A., & Loewenstein, J. (2020). *The craft of creativity.* Stanford University Press.

Finke, R. A., Ward, T. B., & Smith, S. M. (1992). *Creative cognition: Theory, research, and applications.* The MIT Press.

Fisch, F. (1994). Eine Methode zur Analyse von Interaktionsprozessen beim Problemlösen in Gruppen [A method for the analysis of interaction processes during group problem solving]. *Gruppendynamik, 25*, 149–168.

Friard, O., & Gamba, M. (2016). BORIS: A free, versatile open-source event-logging software for video/audio coding and live observations. *Methods in Ecology and Evolution, 7*(11), 1325–1330. https://doi.org/10.1111/2041-210x.12584

Gish, L., & Clausen, C. (2013). The framing of product ideas in the making: A case study of the development of an energy saving pump. *Technology Analysis & Strategic Management, 25*(9), 1085–1101. https://doi.org/10.1080/09537325.2013.832746

Goh, K. T., Goodman, P. S., & Weingart, L. R. (2013). Team innovation processes: An examination of activity cycles in creative project teams. *Small Group Research, 44*(2), 159–194. https://doi.org/10.1177/1046496413483326

Kauffeld, S., Lehmann-Willenbrock, N., & Meinecke, A. L. (2018). The advanced interaction analysis for teams (act4teams) coding scheme. In E. Brauner, M. Boos, & M. Kolbe (Eds.), *The Cambridge handbook of group interaction analysis* (pp. 422–431). Cambridge University Press.

Klonek, F. E., Meinecke, A. L., Hay, G., & Parker, S. K. (2020). Capturing team dynamics in the wild: The communication analysis tool. *Small Group Research, 51*(3), 303–341. https://doi.org/10.1177/1046496420904126

Kozlowski, S. W. J., & Bell, B. S. (2008). Team learning, development, and adaptation. In V. I. Sessa & M. London (Eds.), Work Group learning (pp. 15–44). Mahwah, NJ: LEA.

Leonardi, P. M. (2011). Innovation blindness: Culture, frames, and cross-boundary problem construction in the development of New Technology Concepts. *Organization Science, 22*(2), 347–369. https://doi.org/10.1287/orsc.1100.0529

Leone, S., Japp, P., & Reiter-Palmon, R. (2023). The emergence of problem construction at the team-level. *Small Group Research, 54*(5), 639–670. https://doi.org/10.1177/10464964231152877

Mangold. (2014). *INTERACT user manual*. www.mangold-international.com

Merrifield, P. R., Guilford, J. P., Christensen, P. R., & Frick, J. W. (1962). The role of intellectual factors in problem solving. *Psychological Monographs: General and Applied, 76*(10), 1–21. https://doi.org/10.1037/h0093850

Mumford, M. D. (Ed.). (2011). *Handbook of organizational creativity*. Academic Press.

Mumford, M. D., Mobley, M. I., Reiter-Palmon, R., Uhlman, C. E., & Doares, L. M. (1991). Process analytic models of creative capacities. *Creativity Research Journal, 4*(2), 91–122. https://doi.org/10.1080/10400419109534380

Mumford, M. D., Scott, G. M., Gaddis, B., & Strange, J. M. (2002). Leading creative people: Orchestrating expertise and relationships. *The Leadership Quarterly, 13*(6), 705–750. https://doi.org/10.1016/s1048-9843(02)00158-3

Osborn, A. F. (1953). *Applied imagination: Principles and procedures for creative problem-solving*. Charles Scribner & Sons.

Perry-Smith, J. E., & Mannucci, P. V. (2017). From creativity to innovation: The Social Network drivers of the four phases of the idea journey. *Academy of Management Review, 42*(1), 53–79. https://doi.org/10.5465/amr.2014.0462

Puccio, G. J., Burnett, C., Acar, S., Yudess, J. A., Holinger, M., & Cabra, J. F. (2018). Creative problem solving in small groups: The effects of creativity training on idea generation, solution creativity, and Leadership Effectiveness. *The Journal of Creative Behavior, 54*(2), 453–471. https://doi.org/10.1002/jocb.381

Reiter-Palmon, R., & Illies, J. J. (2004). Leadership and creativity: Understanding leadership from a creative problem-solving perspective. *The Leadership Quarterly, 15*(1), 55–77. https://doi.org/10.1016/j.leaqua.2003.12.005

Reiter-Palmon, R., Mumford, M. D., O'Connor Boes, J., & Runco, M. A. (1997). Problem construction and creativity: The role of ability, cue consistency, and active processing. *Creativity Research Journal, 10*(1), 9–23. https://doi.org/10.1207/s15326934crj1001_2

Reiter-Palmon, R., Herman*, A.E., & Yammarino, F. (2008). Creativity and cognitive processes: A multi-level linkage between individual and team cognition. In M. D. Mumford, S. T. Hunter, and K. E. Bedell-Avers (Eds.), Multi-level Issues in Creativity and Innovation, (vol. 7, pp. 203-267). Bingley, UK: JAI Press.

Reiter-Palmon, R., & Robinson, E. J. (2009). Problem identification and construction: What do we know, what is the future? *Psychology of Aesthetics, Creativity, and the Arts, 3*(1), 43–47. https://doi.org/10.1037/a0014629

Reiter-Palmon, R., Sinha, T., Gevers, J., Odobez, J.-M., & Volpe, G. (2017). Theories and models of teams and groups. *Small Group Research, 48*(5), 544–567. https://doi.org/10.1177/1046496417722841

Reiter-Palmon, R., & Murugavel, V. (2018). The effect of problem construction on team process and creativity. *Frontiers in Psychology*. https://doi.org/10.3389/fpsyg.2018.02098

Reiter-Palmon, R., & Paulus, P. (2019). Cognitive and social processes in team creativity. In M. D. Mumford (Ed.), Creativity and Innovation in Organizations (pp. 161–190). Taylor and Francis.

Rostan, S. M. (1997). A study of young artists: The development of artistic talent and creativity. *Creativity Research Journal, 10*(2–3), 175–192. https://doi.org/10.1080/10400419.1997.9651216

Runco, M. A. (2010). Divergent thinking, creativity, and ideation. In *The Cambridge handbook of creativity* (Vol. 413, p. 446).

Silverman, B. G. (1985). The use of analogs in the innovation process: A software engineering protocol analysis. *IEEE Transactions on Systems, Man, and Cybernetics, 15*, 30–44.

Sternberg, R. J. (1988). A three-facet model of creativity. In R. J. Sternberg (Ed.), *The nature of creativity: Contemporary psychological perspectives* (pp. 125–147). Cambridge University Press.

Ward, T. B., Smith, S. M., & Finke, R. A. (1999). Creative cognition. In R. J. Sternberg (Ed.), *Handbook of creativity* (pp. 189–212). Cambridge University Press.

Weingart, L. R., Cronin, M. A., Houser, C. J. S., Cagan, J., & Vogel, C. M. (2005). Functional diversity and conflict in cross-functional product development teams: Considering representational gap and task characteristics. In L. L. Neider & C. Schreishman (Eds.), *Research in management* (Vol. 4, pp. 89–110). IAP.

Weingart, L. R., Todorova, G., & Cronin, M. A. (2010). Task conflict, problem-solving, and yielding: Effects on cognition and performance in functionally diverse innovation teams. *Negotiation and Conflict Management Research, 3*(4), 312–337. https://doi.org/10.1111/j.1750-4716.2010.00063.x

World Economic Forum. (2020, October 12). *These are the top 10 job skills of tomorrow – And how long it takes to learn them.* https://www.weforum.org/stories/2020/10/top-10-work-skills-of-tomorrow-how-long-it-takes-to-learn-them/

World Economic Forum. (2025, January 7). *Future of jobs report 2025: 78 million new job opportunities by 2030 but urgent upskilling needed to prepare workforces.* https://www.weforum.org/press/2025/01/future-of-jobs-report-2025-78-million-new-job-opportunities-by-2030-but-urgent-upskilling-needed-to-prepare-workforces/

CHAPTER 5

Scientific Collaboration as Multilevel Interaction Processes: Dialectic Knowledge Creation Across Teams, Communities, and Society

Yasunori Yamanouchi, Sachiko Kiyokawa, and Takeshi Okada

Multilevel Collaboration in Science

Beyond the Myth of the Lone Scientist

This chapter focuses on scientific endeavors as a representative domain of collaborative creative problem-solving. While there may be a belief that scientists work individually to achieve major discoveries, actual scientific activities are highly collaborative, involving discussions in laboratories and at academic conferences. Here, we focus on the roles played by collaboration at multiple levels in scientific endeavors and their interrelationships.

Y. Yamanouchi (✉)
Tohoku University, Institute for Excellence in Higher Education, Sendai, Japan
e-mail: yasunori.yamanouchi.d2@tohoku.ac.jp

S. Kiyokawa • T. Okada
The University of Tokyo, Graduate School of Education, Tokyo, Japan
e-mail: kiyo@p.u-tokyo.ac.jp; okadatak@p.u-tokyo.ac.jp

N. Bonnardel (ed.), *Collaborations in Creativity*, The Seven C's of Creativity, https://doi.org/10.1007/978-3-032-07753-0_5

Csikszentmihalyi (2014) conceptualized creativity not as a trait inherent in individuals but as a phenomenon emerging from the systemic interaction among three components: ***person***, ***field***, and ***domain***. In this model, the person generates novel ideas or variations; the field, comprised of gatekeepers and experts, evaluates these variations to determine their worthiness of retention; and the domain serves as a repository and transmission channel, preserving selected innovations for future generations. Creativity, thus, is the result of dynamic interrelations among these three components.

When viewed through this systems lens, creativity in science is typically characterized by individuals who demonstrate analytic, logical, and critical thinking and who persistently engage in skill acquisition. The field, in this case, is primarily constructed by a peer review system that upholds principles of reproducibility, objectivity, empirical validation, and logical coherence. The domain, as a body of disciplinary knowledge, is socially esteemed, structurally cumulative and, while requiring expertise for access, provides clearly defined pathways for mastery.

This chapter similarly adopts a systemic perspective on scientific creativity. However, it shifts the analytical focus from the domain to the collaborative dynamics between the person who generates ideas and the field that evaluates them. Specifically, it examines idea selection processes within collaborative research teams as well as idea generation in the context of peer review communication. Consequently, the paper does not strictly adhere to the dichotomy of idea generation at the individual level and idea selection at the field level. Furthermore, in defining the field responsible for evaluating ideas in a given domain, the analysis expands beyond narrowly defined academic communities to include interdisciplinary consortia and broader societal stakeholders. This approach reflects recent developments in science-society relations and the expansion of Open Science frameworks.

This chapter first provides evidence that collaboration in a small group has advantages in problem solving, including scientific discoveries. Then, based on field research, we extend the scope to larger-scale collaborations, such as collaboration among scientists within academic communities and between scientists and society. Finally, we provide a perspective for understanding scientific endeavors as collaborative processes at three levels, that is, in the real world: teams, academic communities, and society.

Although there are various approaches to understanding scientific endeavors, including sociology, cultural anthropology, philosophy, psychology, cognitive science, and cultural studies, we primarily adopt approaches from cognitive science and psychology, analyzing individual researchers' behaviors, utterances, and thought processes. Various systems used in scientific research, such as the peer review system, are examined from the perspective of how they influence individual researchers' behaviors and cognitive processes. However, if necessary, concepts from other approaches are also incorporated to advance our discussion.

Three Levels of Collaboration: Teams, Communities, and Society

As symbolized by the phrase "standing on the shoulders of giants," scientific endeavor is inherently collaborative. Scientific endeavors are not conducted by lone scientists but through collaboration. According to Dunbar (1997), many great scientific discoveries in molecular biology were not made by lone scientists but were accomplished by groups. Furthermore, as Okada (1999) and Thagard (1997) have pointed out, collaborative research activities are not limited to natural sciences like molecular biology but are equally prevalent in academic disciplines such as the social sciences. Additionally, as of the late 1990s, there has been an increasing trend in collaboration between researchers within the same field and in interdisciplinary research, making interdisciplinary collaboration no longer a special form of research (Schunn et al., 1998).

Collaboration is not limited to a laboratory or research team. While brilliant individual scientists have certainly played crucial roles in scientific advancement, the development of modern science was achieved mainly by creating a system that allowed groups of ordinary people to generate and accumulate knowledge systematically. As Nielsen states, "In the early 17th century, achieving even modest scientific progress required extraordinary genius. However, by the late 17th century, after early modern scientists had established methods of scientific discovery, such scientific advances became commonplace and were achieved by many scientists. Work that once required genius became routine, leading to explosive scientific development" (Nielsen, 2013, p. 10).

Generation and Elaboration of Knowledge Through Dialectic in Collaboration with Multilevel Interactions

How has science developed through collaboration? Science can be seen as knowledge growth. Campbell's (1960) Blind Variation and Selective Retention (BVSR) theory describes knowledge growth as an iterative process involving blind variation—where new ideas are generated without knowing in advance which will succeed—and selective retention, where successful ideas are preserved and serve as a basis for further exploration. Although in science, ideas are not randomly generated, and rarely are specific ideas selected and retained in their original form, it is useful to regard science in terms of two stages: *generation* and *elaboration*.

Collaboration may play a role in the first stage of idea generation. Specifically, collaboration may increase the quantity and quality of ideas. The more people who generate ideas in collaboration, the more ideas will be generated in a team, and the more diverse they will be. As a result, the chances of a good idea being among them may increase. Collaboration may also play a role in the second stage of elaboration. Specifically, generated ideas typically undergo scrutiny and refinement through criticism and supplementary research. Having a variety of perspectives allows us to select the most promising ideas from among many. Furthermore, mutual criticism may help refine promising ideas into more robust ones. These refined ideas can be documented and passed on to future generations for further verification (Todayama, 2020).

In this chapter, we describe the process of science that develops through collaboration as dialectic. Dialectic is a philosophical method that integrates opposing elements to produce a higher-level understanding. This method aims to gain deeper insights by facilitating dialogue between different positions and opinions. This process is explained in three stages: *thesis*, *antithesis*, and *synthesis*, describing the phenomenon where opposing views generate new understanding. Idea generation primarily corresponds with the stage of producing the thesis and antithesis, while idea evaluation primarily corresponds with the synthesis stage.

We conceptualize scientific activity as a fractal structure comprising: (1) idea generation and (2) idea elaboration through collaboration at three levels: *team*, *community*, and *society* (See Fig. 5.1). Team-level collaboration is among groups working on a research project, typically exemplified by collaboration within laboratories or research projects. Community-level collaboration involves selecting, refining, and preserving multiple ideas produced at the team level, with typical examples including peer review in academic communities and discussions at conferences. Societal-level collaboration refers to both intentional and unintentional collaboration that occurs among various stakeholders within society. Intentional collaboration encompasses events incorporating citizens' voices in science and technology policy decisions (e.g., World Wide Views, n.d.).

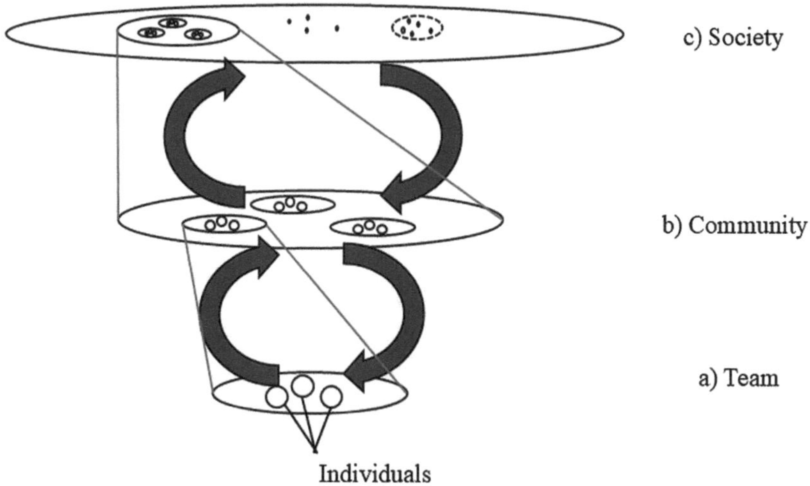

Fig. 5.1 Collaborations at three levels in scientific activities

Unintentional collaboration describes instances where various citizen responses emerge in reaction to scientific reporting, subsequently influencing scientific endeavors and science and technology policies. The topics of this collaboration are diverse, including knowledge emerging at the academic community level, the processes of generating knowledge and technology, and their practical applications. Representative examples include science journalism, science communication, and open science, which continue to diversify with the development of communication technologies, particularly the Internet.

When applying this process to scientific endeavors at the team level, for example when a team conducts research in a laboratory, individual researchers put forward hypotheses, examine multiple research procedures, and consider various interpretations of data. The research progresses as various ideas presented at each stage are examined and refined, ultimately being developed into papers. This process does not proceed in a single direction from idea generation to idea examination and refinement; sometimes new ideas emerge by examining existing ones. As a product of such team collaboration, papers are presented to the scientific community.

Next, collaboration at the community level starts. Specifically, the generation of diverse ideas occurs through individual researchers and research teams in the form of submitted papers. The academic community then carries out the examination and refinement of ideas through peer review and journal publication. It is important to note that when we refer to being "carried out by the research community," we mean the work can be broken down into individual activities: reviewers' comments, authors' revisions based on these comments, and reviewers' decisions on acceptance for publication. Through these interconnected individual activities, promising ideas are published in papers and textbooks, becoming scientific knowledge shared by humanity. This shared knowledge then becomes input for individual researchers, influencing idea generation at the team level. In this way, scientific endeavors advance through idea generation and idea examination/refinement occurring in a fractal-like structure at each level, with these levels continuously interacting.

Collaboration also occurs beyond research teams and academic communities at the societal level. As scientific progress increasingly impacts society, public interest in science has grown. Modern communication tools enable various information about researchers to reach society directly through not only scientific reporting but also researchers', laboratories', and academic societies' websites and social media, which includes both research findings and views on society and politics. Conversely, citizens' opinions about science and research more easily reach researchers, directly and indirectly, influencing idea generation and refinement processes.

In the following sections, we look in detail at the role played by collaboration at team, community, and society levels in science. Finally, we show how collaboration in future science should work on three levels.

Collaboration at Team Level

Interaction with Others Hinders Idea Generation

One of the representative research areas examining the effects of collaboration on problem solving is the empirical investigation of brainstorming effectiveness, a group idea generation technique developed by Osborn (1957) (for reviews, see Homma, 1996; Mullen et al., 1991; and see also in this book Chap. 3 by Girandola & Bonnardel). This brainstorming technique involves generating ideas in groups subject to four rules: (1) generate as many ideas as possible, (2) generate free and uninhibited ideas without being constrained by facts or fixed notions, (3) do not evaluate or criticize ideas that are presented, and (4) exchange ideas to develop and refine them. Following these rules, interaction with others was believed to produce more ideas than working individually. However, empirical results have contradicted this expectation. Interaction with others fails to increase the number of ideas and decreases it. This phenomenon is called *process loss*, where the performance of interacting groups is lower than nominal groups, where results are merely aggregated without interaction.

Similar to brainstorming research, group remembering studies have consistently demonstrated that interactions with others can interfere with remembering (for a review, see Marion & Thorley, 2016). This field examines the comparison between individual recall (individual condition) and group recall (group condition) after learning the same list, and investigates the effects of interaction while controlling for the number of participants involved in the task. This procedure compares the performances of interacting groups and nominal groups the same as in brainstorming research. Studies using various materials have shown that while memory performance in group conditions is better than that in individual conditions, it is worse than that in nominal groups where results are aggregated (stories: Andersson & Ronnberg, 1996; words: Basden et al., 1997: word lists; meaningful images, words, and episodes: Weldon & Bellinger, 1997). This phenomenon, termed *collaborative inhibition*, is interpreted as interference in recall caused by interaction with others and can be considered analogous to process loss in brainstorming research.

Interaction with Others Facilitates Idea Elaboration

There is evidence that interaction with others in a team has a facilitative effect during the idea elaboration stage, while interaction with others has been shown to have a disruptive effect during the idea generation stage mentioned above. Okada and Simon (1997) conducted an experiment to examine the effects of collaboration on scientific discovery. Of the 27 participants, 18 (9 pairs) were assigned to the pair condition, while nine were assigned to the individual condition. The task employed was the Simulated Molecular Genetics Laboratory (SMG Laboratory), a molecular genetics task developed by Dunbar and colleagues (Dunbar, 1993; Dunbar & Schunn, 1990). Although simplified, this task effectively simulated an actual collaborative discovery process. Initially, task performance was compared between individual and pair conditions. The results demonstrated higher performance in the pair condition than the individual condition, confirming that pairing facilitated scientific discovery. Subsequently, a detailed protocol analysis was conducted to examine the factors contributing to these performance differences. The analysis revealed that pairs outperformed individuals because they more frequently engaged in *explanatory activities*, including hypothesis generation, discussions of competing hypotheses, and hypothesis testing to co-construct new knowledge based on experimental data as evidence.

Collaboration at the Community Level

Characteristics of Collaboration at the Community Level: The Tension Between Co-creation and Competition

Community members are colleagues who share research paradigms and co-create knowledge. However, they simultaneously function as rivals competing to reach the truth more quickly and accurately. Since research demands originality, each researcher and research team maintains their unique assertions, though to varying degrees. These assertions may conflict with one another, and when such conflicts are resolved, they may result in different stakes for each member. This tension characterizes the production, examination, and refinement of ideas at the community level.

Typical examples of community-level collaboration include peer review in academic societies and discussions at annual conferences. In the following, section "Peer Review: The Evolution of Mechanisms Supporting Reliability" provides an overview of peer review as a representative example of community-level collaboration. Section "Mutual Criticism in Symposiums: Learning from Practice" then presents a case study of how collaborative efforts at academic symposia led to the development of research evaluation criteria (Yamanouchi & Okada, 2003a, 2003b, 2003c).

Peer Review: The Evolution of Mechanisms Supporting Reliability

While peer review has a long tradition, it remains relatively young as an institutional practice. Although activities leading to modern peer review have existed for 350 years, its formal implementation in journals did not begin until the 1960s (Ware, 2017). For instance, in 1973 *Nature* initiated a systematic external peer review for all submitted papers in 1973 (Baldwin, 2015). In recent years, publication formats have diversified with online journals and preprints, and peer review continues to evolve accordingly.

Despite these ongoing changes, peer review retains a central role in the quality control of ideas. It selects promising ideas through accept/reject decisions for journal publication, ensuring that published ideas and research reliability remain above a certain standard. The digitization of scholarly information dissemination has inundated researchers with an overwhelming volume of information, prompting the development of various alternative metrics for assessing the reliability of academic articles, such as impact factors. Nevertheless, peer review remains a cornerstone for evaluating the reliability of information, and its influence has not only endured but has grown stronger over time (Nicholas, 2015).

However, peer review faces numerous challenges and criticisms. These include costs and time requirements, lack of reliability, effectiveness, fairness, insufficient transparency, and potential suppression of innovative research (Kelly et al., 2014; Sato, 2016). Securing qualified reviewers has become increasingly difficult as publication volumes increase and research fields become specialized. Additionally, intensifying competition among researchers limits the time they can dedicate to reviewing. Against this background, establishing a sustainable peer review system remains a significant challenge.

Collaboration in peer review corresponds with examining and refining ideas rather than generating them, as ideas are already submitted in paper form. Previous research suggests reviewer feedback provides opportunities to reconsider and modify argument structure, knowledge claims, presentation style, and word choice (Cowley, 2015). Reviewers generally contribute to creating papers acceptable to the broader academic community by offering perspectives different from those of team members. However, due to limited studies, the collaborative process in peer review remains unclear. Criticized for its opacity, peer review typically occurs in private, and even when review comments and responses are published, the underlying thought processes remain hidden.

In response to these criticisms, various peer review approaches are being explored, potentially creating new forms of collaboration (Wiley, n.d.). For example, open peer review reveals both authors and reviewers, and some academic journals publish review results alongside final papers with the consent of both parties. Open peer review is expected to motivate reviewers to conduct more careful and higher-quality reviews than anonymous reviews, thereby improving the quality of collaborative idea elaboration.

For example, the *Journal of Interactive Media in Education* (JIME) has implemented a dialogic peer review process aimed at fostering discourse among stakeholders from diverse backgrounds—including researchers, educators, system designers, and policy makers. Through the integration of purpose-built, document-centered discourse interfaces and a shift from anonymous critique to collaborative editorial mediation, the role of reviewers has evolved from that of mere evaluators to that of co-editors. This development represents a novel phenomenon within the scholarly publishing system, highlighting the transformative potential of open peer commentary in mediating knowledge construction and facilitating interdisciplinary collaboration (Chevalier et al., 2000; Sumner et al., 2000).

Furthermore, collaborative peer review implements a team-based approach to manuscript evaluation where multiple reviewers work together to review a paper, discuss their opinions, and submit a unified report, or where one or more reviewers collaborate with authors to improve the manuscript to a publishable standard. In these approaches, the benefits of collaboration are likely to be more pronounced.

Mutual Criticism in Symposiums: Learning from Practice

Formation of Validation Boundaries and Methodological Principles

Through peer review, from among multiple ideas, promising ideas are examined and refined. However, how are the criteria for examining these ideas formed in the first place?

Fujigaki (1998) defines the boundary of knowledge judgment in a journal as the validation boundary and identifies three key characteristics:

- Validation boundaries do not exist from the beginning but are formed through the accumulation of reviewers' acceptance and rejection decisions.
- The validation boundary of the latest issue of a specialized journal is actively being created, continuously being defined, rewritten, and updated in real time.
- It is formed under conditions ideal for the scientific community rather than conditions necessary for solving public problems.

Furthermore, this boundary is not explicitly documented and researchers typically remain unconscious of it until they encounter people with other boundaries or lay people in the field, at which point it becomes apparent.

Validation boundaries are generally formed through community-level collaboration, such as accumulating long-term peer reviews and aligning perspectives at academic conferences. However, due to the above-mentioned characteristics, the formation process is difficult to document. Since validation boundaries are constantly updated, obtaining comprehensive data on the formation process is challenging. Consequently, describing the formation process of validation boundaries presents significant difficulties.

However, it is not impossible. Latour (1987) presents a methodological principle for revealing "how scientists achieve something" rather than what scientists have achieved, stating that we should "study science in the making, not ready-made science and technology. To do this, we must either arrive before facts and machines become black-boxed or follow the controversies that reopen them." Following this principle, we should trace the controversies that reopen validation boundaries.

One such controversy occurs when conventional validity boundaries are called into question due to the discovery of research misconduct, such as data fabrication. The relevant scientific community must urgently establish new boundaries capable of rejecting fabricated data that could not be rejected under the previous validity boundaries. Consequently, intensive communication regarding validity boundaries is expected to occur. The emergent formation of such validity boundaries differs from the usual time-consuming formation process. However, even in an emergency, since these boundaries will be applied to subsequent verification and research, the process is not entirely different but rather involves concentrated discussion of the essential core elements typically present in the normal formation process. It is necessary to be careful about the differences between emergency and normal situations. However, as in the debates that have taken place in various fields when reproducibility has been called into question, it is certainly true that emergent revision of standards is an actual part of scientific practice.

Case Overview

In November 2000, the Japanese archaeological community was shaken by the Paleolithic excavation fabrication case, exposed through an investigative report by the Mainichi Shimbun (2000a, November 5). The newspaper captured footage of an individual burying stone tools at excavation sites. This individual, known as "God's Hand," had discovered numerous stone tools considered to be among Japan's oldest, including some from nationally designated historic sites and archaeological sites featured in textbooks. When interviewed, he claimed that fabrications occurred at only two archaeological sites where evidence had been found, though there was no guarantee that this statement was true. Following the exposure of these fabrications, he experienced severe physical and mental health deterioration, resulting in hospitalization, and medical professionals determined that further questioning would not be possible. Given these circumstances, the Japanese Archaeological Association decided to conduct a comprehensive investigation to fulfill its social responsibility. The Japanese Archaeological Association subsequently examined approximately 200 sites and 3,000 stone tools, announcing in May 2003 that large-scale fabrication had occurred for over 25 years. The beginning of Japan's Paleolithic culture, previously thought to date back 700,000 years, was revised to 30,000 years ago.

A crucial point here is that when the incident came to light in 2000, it was not archaeologists who detected the fabrication, but rather a newspaper that broke the story. In contrast, the full extent of the fabrication revealed in 2003 was uncovered through academic verification by archaeologists. The archaeological community initially lacked a methodology to detect fabrications and developed new methodological approaches in less than three years.

Yamanouchi and Okada (2003a, 2003b) focused on discussions at two symposiums held within three months after the incident's discovery. These symposiums not only attracted public attention, with their proceedings later published, but also served as the first venues where three new criteria for subsequent verification work were proposed and discussed. Specifically, these were gajiri (unnatural recent damage on stone tool surfaces), limonite (artificial metallic marks from agricultural tools on stone tool surfaces), and pressure flaking (traces of pressure-induced flaking and processing). Approximately one third of the discussion was devoted to these criteria (Yamanouchi & Okada, 2003a, p. 423).

Gajiri and limonite are commonly found in stone tool materials collected from cultivated topsoil in more recent periods. In contrast, Paleolithic artifacts are contained in layers deep below the surface, beyond the reach of agricultural tools, so these marks cannot appear except during excavation work. Typically, the oldest Paleolithic artifacts are carefully excavated using wooden spatulas to avoid damage. These marks would only appear if careless excavation used metal trowels with forceful contact. Even in such cases, excavators switch to wooden spatulas once contact is made, preventing multiple marks from occurring. Therefore, if multiple gajiri or limonite marks are found on a single stone tool or multiple tools from the same site exhibit these characteristics, they were likely in cultivated soil and probably not excavated from the reported stratigraphic layer. Furthermore, since cultivated soil generally contains mixed layers from different periods, stone tools excavated from such soil cannot be dated based on their stratigraphic position, making it difficult to treat them as authentic data.

The pressure flaking technique, outside Japan, is only documented from the Upper Paleolithic period onward. However, it had been found in Japan's Early and Middle Paleolithic layers. Before the fabrication was exposed, some claimed that early Japanese humans were exceptional or that similar evidence had not yet been discovered elsewhere. After the fabrication's exposure, pressure-flaked tools in Early and Middle Paleolithic contexts were recognized as anomalous based on worldwide archaeological evidence.

In this case, some symposium participants presented these criteria during their presentations, though the original development of these ideas remains unclear. The following sections examine how other participants responded to these proposals, particularly on counter-arguments (criticisms and alternative hypotheses), and how these challenges helped refine the original criteria.

While a "when in doubt, exclude it" approach might seem prudent to avoid misidentifying fraudulent artifacts as genuine, the rarity of Early and Middle Paleolithic artifacts makes misidentifying genuine artifacts equally problematic. Historical data cannot be experimentally reproduced. Furthermore, these criteria would affect future Paleolithic archaeological research, and overly strict standards could impede scientific progress. This context explains why the development of fabrication detection criteria proceeded with such careful consideration.

Generation of Counterarguments

As a result of resolving conflicts over a specific claim through collaboration at a community level, community members have different stakes. For example, if a stone tool is determined to be a forgery, the member who discovered or reported on it in an academic paper will lose their scholarly achievement. How do such personal interests affect the generation of counterarguments?

Yamanouchi and Okada (2003c) analyzed the above dataset by dividing speakers based on whether they were involved in the excavation suspected of forgery. They examined the presence of counterarguments to the proposal of new criteria. The results showed that out of 422 relevant statements regarding the three criteria, there were 57 counterarguments. These counterarguments were made by 10 out of 16 symposium participants. Seven of these 10 individuals were involved in the excavation, while three were not. This result was statistically significant at the 5% level and was consistent with the interpretation that researchers involved in the excavation were influenced by their motivation to protect their scholarly achievements.

Examination and Refinement Through Mutual Criticism

How did these objections influence the formation of criteria? If personal goals influenced them, they could potentially damage collaboration. To analyze the content of objections, Yamanouchi and Okada (2003a) assumed that if the formation of criteria for determining Early Paleolithic artifacts were conducted rationally, the discussion would follow modus ponens. That is, from "Proposition 1: $P \Rightarrow \neg Q$" (if characteristic X is observed, it is not Early Paleolithic) and "Proposition 2: P" (characteristic X is observed), the conclusion $\neg Q$ (it is not Early Paleolithic) is derived.

Arguments concerning Proposition 1 question the validity of the criteria themselves, while arguments concerning Proposition 2 question the validity of the procedures for applying these criteria (Yamanouchi & Okada, 2003a, p. 430). When gajiri (plow marks), limonite, or pressure flaking is substituted for characteristic X, if Propositions 1 and 2 are accepted as true, it can be logically determined that "it is not Early Paleolithic." Conversely, if either Proposition 1 or 2 is false or impossible to determine when substituting characteristic X, that characteristic would be excluded from consideration as a new criterion.

Content analysis of 57 objection statements revealed 10 objections to Proposition 1, 38 objections to Proposition 2, and 9 other objections. These results indicated that these two propositions were the main points of contention. Furthermore, significant variations emerged when comparing the number of objections to Propositions 1 and 2 across the characteristics of gajiri, limonite, and pressure flaking. Consequently, Yamanouchi and Okada (2003b) described and compared the formation processes of these three criteria within this framework. For example, regarding gajiri, the discussion proceeded as follows: Concerning Proposition 1, when gajiri was defined as "damage caused by cultivation," it implied that the stone tools were contained within cultivated soil. If they were found in cultivated soil near the surface, this necessarily meant they were not in deeper Early Paleolithic strata. Therefore, they could not be considered Early Paleolithic artifacts. No objections were raised to this logic.

However, regarding Proposition 2, two main issues were raised: "judgment criteria" and "observation methodology." The judgment criteria issue concerned determining the cause of damage to stone tools. Since stone tools might have existed for hundreds of thousands of years, numerous possibilities for physical contact could have occurred during that time, potentially resulting in various forms of damage. Similarly, cultivation-related damage could take various forms, raising questions about whether the cause could be determined from the shape alone. The methodological issue addressed what observation methods were appropriate. Some advocated visual observation, emphasizing quick verification by multiple people at a low cost, while others promoted microscopic observation, emphasizing rigorous verification by experts. These proposals were incorporated into subsequent verifications, leading to the establishment of more robust verification methods.

In the gajiri discussion, while this type of damage on stone tools was known as "gajiri" in excavation site terminology, it had not been academically recognized, and its morphology and causes had not been studied systematically. After the incident came to light, it became a crucial indicator for determining the stratum containing the stone tools, leading to questions about judgment criteria and observation methods; as the field term *gajiri* transitioned into academic terminology, judgment criteria and observation methods were refined. This refinement process was effectively driven by the objections raised during the discussion.

Coexistence of Interests and Rationality: Emergent Phenomena at the Community Level

In this case study, while individual researchers appeared to be influenced by personal goals in their criticism and refutation of opposing claims and their foundations, the overall discussion led to the refinement of evaluation criteria and observational methods through repeated cycles of claims, rebuttals, and counter-rebuttals.

Iseda (2001) indicates that the future direction of the sociology of scientific knowledge should explain "how a scientific community composed of scientists (who are influenced by personal goals and biases) can, as a whole, continue to form a consensus that can be explained through a rationalistic view of science."

This case demonstrated two key points: (1) The influence of personal goals promotes the discovery of problems in opposing claims and the proposal of competing hypotheses. (2) Through mutual criticism, such problem identification and competing hypotheses lead to the refinement of concepts and observational methods.

The crucial insight here is that a valid rebuttal can contribute to constructive discussion regardless of the underlying intention; even when individual researchers attempt to falsify claims to protect their achievements, the community exhibits falsificationist behavior. Because of personal goals, opposing claims undergo rigorous scrutiny, leading to problem discovery and the generation of competing hypotheses. The discussion becomes more sophisticated as researchers attempt to provide more reliable explanations in response to such rebuttals. Thus, the influence of personal goals on individual researchers can coexist with rational consensus building at the community level. This multilevel perspective helps explain the emergent phenomenon whereby researchers, though influenced by personal interests, behave rationally as a collective.

Collaboration between researchers with opposing theories has recently gained attention as "adversarial collaboration" (Clark et al., 2022). This approach aims to generate new insights through joint experiments by opposing researchers. However, while this study showed common interests and cooperation rules emerging naturally through societal-level collaboration, adversarial collaboration explicitly defines these parameters. Similarly, joint fact-finding offers another example (Ozawa & Susskind, 1985), where stakeholders with conflicting interests work to confirm common facts. While traditionally used in policy domains, this approach will likely become increasingly important across various fields in the post-truth era.

However, such emergent behavior is not universal and appears only under specific conditions (Curhan et al., 2021). Indeed, even before the fabrication was discovered, doubts about Early Paleolithic findings had been raised, but these doubts did not lead to constructive discussion.

Here, we face the following question: under what conditions does emergence occur? One key condition appears to lie in societal-level collaboration.

Collaboration at the Societal Level

Characteristics of Collaboration at the Societal Level: The Expansion of Discussion Generated by Diverse Perspectives

Members participating in societal-level collaboration are not limited to specific sectors and are often anonymous, making collaboration among a large, unspecified number of people the primary focus. While interest in topics is shared, it remains unclear what else is shared and to what extent. A distinctive feature is that the diversity of participants naturally leads to a multifaceted examination of issues.

As points of intersection between science and society increase, venues for collaboration are expanding. These collaborations take various forms, including science journalism, science communication, open science, and industry-government-academia-society partnership projects, depending on their purposes (such as information exchange, research, and social problem-solving), the design of collaborative spaces, and the participants involved.

In this chapter, as examples of societal-level collaboration, we introduce discussions from electronic bulletin boards and science journalism related to the aforementioned archaeological case based on Yamanouchi (2017). Through these examples, we examine the impact of societal-level collaboration on idea generation and the examination and refinement of ideas at the community level.

Collaboration on Electronic Bulletin Boards: New Questions Emerging from Dialogue with Society

The year 2000, when the fabrication was discovered, coincided with the emergence of Web 2.0, and discussions on electronic bulletin boards (BBS, Bulletin Board System) were particularly active. The Japanese Archaeological Association (2003) indicated that regarding the issue of fabrication, a network of interested individuals has continued to collect information through internet-mediated channels (p. 551). Through online information exchange, it has become generally apparent which parties possess specific information on the matter. Here, we discuss idea generation through online collaboration while examining the discussions about fatty acid analysis that took place on these bulletin boards.

Fatty Acid Analysis

A debate emerged on electronic bulletin boards regarding the validity of fatty acid analysis as a scientific method. At the Babadan A site, one of the verification targets, it was claimed that fatty acids from animals (Naumann's elephant) that had been butchered using stone tools 200,000 years ago were detected through fatty acid analysis. Even after the fabrication was discovered, some Paleolithic researchers continued to cite this result as valid evidence against fabrication. Questions about this were raised on the electronic bulletin board.

Subsequently, academic presentations and papers based on these online discussions were presented in formal academic venues. While the primary authors were not archaeologists who had participated in the bulletin board discussions, archaeologists who had participated were included as coauthors. This criticism was covered in both newspapers and the final report, contributing to reassessing the relationship between natural science and archaeology.

The bulletin board discussions revealed different perspectives between archaeologists and specialists from other fields regarding "what should be scientifically explained." Traditionally, archaeology focused on identifying animal species from fatty acids, often referring to them as "Naumann's elephant fatty acids" even after the fabrication was discovered. This focus stemmed from the importance of determining which animal species' fatty acids were present to infer Paleolithic living environments.

In contrast, experts from other fields posting on the bulletin boards raised questions about the raw data from the fatty acid analysis and the chemical compound names of the "supposed Naumann's elephant fatty acids." They posed questions previously unasked in archaeology, such as how fatty acids could remain on stone tool surfaces despite vulnerability to cellular autolysis, microbial growth, and oxidative decomposition; what special conditions allowed for fatty acid preservation; and whether experimental results could truly support the conclusion that fatty acids could persist for 200,000 years. These experts viewed the 200,000-year preservation of fatty acids as an extraordinary chemical phenomenon.

This demonstrates how research questions derived from the same phenomenon or data can vary significantly across disciplines. Knowledge differences regarding physicochemical properties were shown to influence the discussion. For example, the question "Could a 5% concentration of palmitoleic acid be detected on a 200,000-year-old stone tool surface?" arose specifically because of palmitoleic acid's physicochemical properties (1. it is liquid at room temperature and therefore leaches out, 2. it readily oxidizes, converting from unstable unsaturated fatty acids to stable saturated fatty acids).

In this case, questions that had never been raised within the archaeological community were discussed through broader societal collaboration, clearly revealing the limitations of fatty acid analysis. These findings represent achievements made possible through collaboration at a societal level.

Journalism's Critique of Science: Questioning Research Culture

Before the fabrication was exposed, critical papers about Japanese Early Paleolithic research (such as Oda & Keally, 1986) had been published but had been systematically ignored. On the other hand, as discussed in section "Collaboration at the Community Level", archaeologists had constructed new standards and observation methods through mutual criticism after the exposure of the fabrication. Why did the archaeological community, which had ignored criticism when the critical papers were published, not ignore it after the fabrication was exposed?

Here, we focus on the role of society surrounding the research community. Journalists, in particular, are considered "watchdogs who always view their 'subjects' with a critical eye, which is their social role and responsibility" (Kovach & Rosenstiel, 2014, p. 28). Science journalists are expected to monitor science with a critical eye. What impact did science journalists have on the archaeological community?

The science journal *Nature* published an article about this incident within two weeks of its discovery (Cyranoski, 2000). However, it focused not on the fabrication but on problems with Japanese research culture. The article "Fake finds reveal critical deficiency" began with: "Last week's disclosure of extensive fraud by one of Japan's leading archaeologists has renewed soul-searching about how much the nation's more prominent scientists are allowed to escape criticism by their peers."

In Japan as well, in a feature article on the fabrication incident, the Mainichi Shimbun (2000b, November 15) criticized the archaeological community for suppressing dissenting opinions and avoiding scholarly debate. Thus, domestically and internationally, the Japanese archaeological community was criticized for lacking critical discourse and subsequent discussion.

In their analysis of society's response to the fabrication incident, the Japanese Archaeological Association (2003, pp. 549–550) addressed criticism regarding the series of investigations beginning with the Zazaragi site. While these investigations had ostensibly resolved the debate over the existence of Early Paleolithic remains, they were criticized for systematically dismissing opposing viewpoints from the outset. The Association acknowledged that the responsibility for failing to rigorously pursue academic questions extended beyond those directly involved in the discoveries to encompass the entire archaeological community. Moreover, they recognized that the discipline of archaeology was being challenged to fundamentally reexamine its academic foundations, emphasizing the need to cultivate a more robust scholarly environment for the field's future development. In addressing academic ethics, they also highlighted the importance of understanding the research systems, institutional culture, and interpersonal dynamics that had allowed such fabrications to persist unchallenged.

In response to these questions, the archaeological community must demonstrate to society its capacity for mutual criticism and self-regulation through actions, not just words. Otherwise, the credibility of the archaeological community and its researchers would be compromised. Specifically, they needed to create open communication forums and make internal discussions visible to society. Indeed, at the symposium discussed in section "Collaboration at the Community Level", opportunities were provided for anyone to examine the suspected fabricated stone tools, discussions were held publicly, and speech records were published. Furthermore, they demonstrated their commitment to addressing criticism seriously, engaging with researchers, journalists, and others from various positions.

When minutes of meetings are made public or discussions are conducted on bulletin boards, inconsistencies between words and actions or contradictory claims are often highlighted. Moreover, criticism directed at others can sometimes return as a boomerang, becoming criticism against oneself. Those who criticize the lack of scientific norms and mutual criticism in such forums must also respond to criticism of their statements. These conditions promote mutually critical and more normative behavior. Society has influenced the examination and refinement of ideas at the community level by raising the normative awareness of the scientific community.

Innovation Born from Diverse Perspectives: Open Science and Citizen Participation

In section "Collaboration on Electronic Bulletin Boards: New Questions Emerging from Dialogue with Society", we examined the role that societal-level collaboration can play in idea generation, and in section "Journalism's Critique of Science: Questioning Research Culture", we explored its role in idea scrutiny and refinement (at the community level). These can be understood through the perspectives of "designed serendipity" and "shared praxis," which Nielsen discussed in the context of open science.

Designed Serendipity

The fatty acid analysis observed on electronic bulletin boards can be understood as "designed serendipity," as Nielsen (2013) described. "Designed serendipity" refers to situations where, when faced with a difficult problem requiring someone with optimal expertise to solve it, rather than relying on chance, the possibility of finding a solution is greatly enhanced by exposing the problem to many participants with diverse expertise.

Not only in open science but also in citizen-participatory technology assessment, which is conducted when introducing advanced science and technology into society, the aggregation and utilization of local knowledge is expected. Social problems cannot be discovered or solved through knowledge from a single specialty alone; rather, knowledge from multiple specialties and citizens' local knowledge must be utilized. Furthermore, since applied science and technology shape society, citizens' opinions must be considered when determining the state and direction of science and society.

In archaeology, the results of fatty acid analysis continued to be considered unquestioned evidence supporting the authenticity of stone tools, even after the discovery of fabrication. However, issues with the fatty acid analysis results and methodology were identified through societal-level collaboration on electronic bulletin boards. The participation of experts from other fields with the necessary capabilities and motivation to examine fatty acid analysis in online discussions enabled investigations that archaeologists could not conduct, including examinations of fatty acids' chemical properties, the validity of extrapolation from experiments, and the reliability of analytical methods.

Thus, societal-level collaboration generated designed serendipity, leading to the discovery of questions from perspectives different from those within the archaeological community, attracting necessary experts to address these difficult problems, and ultimately reaching solutions. If these issues had remained confined to the archaeological community, archaeologists with the necessary knowledge and abilities for verification would not have emerged so quickly. Societal-level collaboration, serving as a matrix for generating designed serendipity, contributed to ensuring the quality of specialized knowledge and techniques within the archaeological community.

Shared Praxis

Scientific endeavor can be viewed as a large-scale collaboration integrated through standards of evidence and reasoning (Nielsen, 2013). For such collaboration to be established, it is said that a "shared praxis"—sharing a certain body of knowledge and technical expertise among group participants—is essential.

Shared praxis in science generally encompasses scientists' codes of conduct and scientific methods characterized by reproducibility, empiricism, objectivity, and mutual criticism. As specialization progresses, this shared praxis is customized into field-specific shared praxis according to each discipline's characteristics. Professional education and laboratory training aim to help practitioners acquire each field's shared praxis. General scientific shared praxis and specialized shared praxis do not necessarily align.

However, they do not become entirely separate entities. General scientific shared praxis is a prerequisite for specialized shared praxis, and its fundamental values are maintained. Conversely, if a field's shared praxis deviates excessively from general scientific shared praxis, that specialty field also deviates from science.

What can collaboration with society bring to the academic community? In this case, it first (1) revealed the divergence between the shared praxis of early and middle Paleolithic research in Japan and general scientific shared praxis. This divergence had been noted for over 20 years. However, it did not become an issue because these observations remained at the community level. As long as discussions remain confined within the community, criticism of the field's shared praxis becomes merely a minority opinion from newcomers or outsiders. For the community majority, the field's shared praxis specificity, like validation boundaries, goes unrecognized, and understandably, criticism is downplayed. However, once collaboration extends to the societal level, triggered by events like the discovery of fabrication, the majority–minority relationship is reversed. Any specialized community becomes a minority from a general science perspective. In societal-level collaboration, deviation from general scientific shared praxis becomes problematic and cannot be ignored. Social trust and standing as a scholar are lost if one insists on specialized shared praxis and is labeled unscientific.

Furthermore, (2) societal-level collaboration effectively corrects divergence. Since community members are often unaware of their field's shared praxis, it is difficult to identify problems, modify them to be socially trusted, and implement changes. At a minimum, evaluating whether modified shared praxis is valid from a general science perspective requires societal-level engagement. Therefore, it is more effective to conduct modifications openly with society. In this case, researchers from other fields and citizens directly pointed out deviations from general scientific shared praxis, and archaeologists responded accordingly. This process encouraged the reexamination of current praxis and problem identification.

Societal-level collaboration also (3) created an environment that encouraged academic communities and individual researchers to act with awareness of general scientific shared praxis. Not only when directly questioned by society but even in scholarly discussions within the community, awareness that these discussions might be viewed by society promotes self-reflection on whether statements are socially acceptable. Individual researchers cannot ignore criticism, even if they might wish to. This is because any deviation from the shared praxis of science in general, when under public scrutiny, would diminish the community's reputation as a whole and call into question the raison d'être of the academic society. Consequently, at an individual level, one's position as a researcher would also be threatened. The latter prevails when weighing research claims against one's position as a researcher. Moreover, because the discussion activated general scientific praxis emphasizing rationality and empiricism, the two propositions discussed in section "Collaboration at the Community Level" became the most important points of contention in rejecting new criteria.

Being conscious of general scientific shared praxis can be a condition for successful collaboration with others having different interests and expertise. Societal-level collaboration promotes the manifestation of shared praxis, prevents specialized fields from excessively deviating from science and, when deviation occurs, provides opportunities to correct problematic shared praxis at the community level while encouraging behavior adhering to general scientific shared praxis at the individual level.

The Future of Collaboration: Knowledge Evolution Through Multilevel Dialogue

Collaboration as Dialectics: From Conflict to Integration

The Balance Between Similarities and Differences

In the case examined in this chapter, three elements made collaboration at multiple levels in science fruitful: (1) conflicting claims arising from divergent interests within the community, (2) designed serendipity that questions and moves toward solutions from perspectives different from the community members, and (3) activation of shared praxis in general science that leads to the emergence of diverse claims and ideas.

Collaboration inherently contains certain contradictions. The existence of shared praxis implies similarities among members. Conversely, designed serendipity arises from differences between members. If the similarity is too high, information does not increase, and collaboration becomes meaningless. It gradually leads to closed communication that excludes external groups, making it impossible to recognize the unique characteristics of the field's shared praxis. However, if differences are too great, there is no common ground for constructive discussion. A balance or appropriate tension between similarity and difference is required. Differences provide the driving force for collaboration, while similarities control this force.

From a dialectical perspective, the "divergent interests" in point 1 motivate scientists on both sides to make their claims more empirical and rational. Team-level collaboration exemplifies the type of collaboration that strengthens individual claims. Research articles present each team's claims to the community empirically and rationally. Since articles require consensus among authors, collaboration among those with similar interests and evaluation criteria has a higher probability of success.

The "diversity of perspectives" in point 2 provides the foundation for generating various antitheses to a single claim. Discussion quality improves through examination from multiple angles and refinement of claims. Societal-level collaboration is effective when introducing these multiple perspectives. Designed serendipity enables access to knowledge not possessed by specific teams or academic communities.

Through "discussion aligned with the shared praxis of general science" in point 3, strong opposing claims and diverse points arising from multiple perspectives are less likely to be ignored or treated disingenuously. Community-level collaboration is ideal for discussions that generate synthesis by understanding and incorporating each team's specialized claims and related diverse points. At the community level, those with opposing specialized claims can engage in direct, equal dialogue, and by fulfilling accountability to society, the academic community can gain social trust. Communities understand general scientific and specialized field shared praxis while maintaining appropriate diversity.

However, if a community becomes isolated from society, there is a concern that it may deviate from the shared praxis of science in general. In this regard, cooperation at the societal level serves as a foundation for generating discussions that lead to synthesis. Furthermore, society may impose constraints on communities. For example, in the case of fabrication discussed in this chapter, it was noted that if decisions about textbook revision were not made by the start of the next academic term, it would create confusion for both educational institutions and students. While multiple hypotheses can coexist within academic communities regarding a particular question, there may be contexts in society where such multiple hypotheses are not acceptable. Thus, society can impose various constraints on academic collaboration, such as demanding early explanations from academic communities.

Collaboration Among Researchers

When viewing collaboration as a dialectic process, individual researchers seek antitheses and participate in various collaborative settings. They produce syntheses that create new ideas by colliding and examining different ideas.

Individual researchers first confront data in their search for antitheses. When data does not fit their ideas, they are forced to revise their concepts. They then seek antitheses from others, beginning with collaboration among laboratory or project members. While personal differences exist, evaluation criteria and interests remain relatively aligned, allowing for criticism in a cooperative atmosphere. They engage in community-level collaboration through papers and conference presentations, receiving specialized antitheses from researchers of different schools of thought, sometimes with conflicting interests. After publication, they present their findings to society, seeking further differences and obtaining multifaceted antitheses. This process objectively improves the quality of individual ideas.

The core of science exists at the community level. Under the current system, ideas originating at individual or societal levels must be proposed, examined, and refined at the community level to gain recognition as academic contributions. While academic communities typically function effectively, they can benefit from societal-level collaboration when experiencing dysfunction (e.g., excessive homogeneity), seeking societal harmony, or requiring new stimuli.

Individuals and teams have inherent limitations in their diversity. Each person's cognitive resources are limited. Therefore, to make nature and society comprehensible, people process information according to their simplified cognitive frameworks, such as confirmation bias. Ignoring information that does not fit within one's cognitive framework and assuming understanding feels satisfying. However, any framework of understanding is subject to revision when viewed objectively. Humanity has not completely understood the world, and the world itself is subject to change. If we remain complacent with a particular framework of understanding, nothing new will emerge. That is why researchers resist the temptation of complacency with the support of others (laboratory members, peer reviewers, and society). They continuously evolve their cognitive frameworks.

Society offers diverse perspectives but has similar limitations and lacks a unifying will. Those seeking to create synthesis form communities sharing certain norms and values. They build synthesis as a community rather than selecting specific individual ideas from many. While selecting from diverse ideas creates economies of scale, science's true work lies in examining and refining even a few hypotheses from multiple angles.

Community members need neither be cordial nor deliberately critical. When researchers make unique claims, differences naturally emerge. Relativism does not generate conflict; identifying and highlighting problems in others' ideas leads to mutual refinement. Meta-analytically, this produces reliable knowledge emergence across the community. Academic conferences and peer review inevitably create opportunities for intellectual collision, benefiting the community. Similarly, outreach activities provide opportunities to respond to societal criticism while leveraging societal wisdom.

Few researchers attend conferences seeking conflict; many seek colleagues and friends. Some research requires collaboration, and researchers naturally desire recognition. However, even among perceived peers, individual differences exist at detailed levels. Science particularly values uniqueness, making differences more pronounced and noticeable. These differences generate questions and competing hypotheses, driving the progress of opposing science. Creating new ideas requires recognizing flaws in existing ones. Since humans are not perfect, no idea is flawless. As humans, social environments, and knowledge evolve, ideas undergo minor disruptions, requiring creativity to bridge emerging gaps.

The Multifaceted Nature of Researcher Identity: Growth Through Collaborative Internalization and Inner Dialogue

A researcher possesses multiple identities. For example, in team-level collaboration, they are members of laboratories and projects; in community-level collaboration, they are members of academic societies and academia; and in societal-level collaboration, they are members of their nation and humanity. Additionally, they maintain identities as family members and workplace members.

Dialogical Self Theory emphasizes that the self is not a singular entity but is composed of multiple selves (Hermans et al., 1992). This theory emphasizes the diversity of the self and its internal dialogue.

These multiple selves shift flexibly according to situations and temporal changes. For instance, when team-level or community-level identities are activated, excluding voices from external groups to facilitate specialized discussions may be considered acceptable. However, when societal-level identities are activated, shared practices and accountability in science become paramount.

These selves are said to influence each other through internal dialogue. For example, when one's research results face criticism, both the individual-level identity (wanting to ignore criticism to protect one's research) and the community-level identity (seeking accurate recognition even if one's research is challenged) may be simultaneously activated. In such cases, simultaneously following different rules may lead to self-contradictory judgments. Through internal dialogue, these contradictions become opportunities for personal growth and change, leading to richer self-awareness. Through this process, each identity is also transformed.

These multiple researcher identities are formed through collaboration at various levels. Having multiple identities enables complex internal interactions, as described above. For example, students internalize their advisors' perspectives through guidance and, through dialogue with these internalized perspectives, examine their research more rigorously (team-level collaboration). Through repeated paper submissions, they internalize reviewers' perspectives (validation boundaries) and learn to write papers acceptable to researchers with diverse, sometimes opposing, viewpoints (community-level collaboration). Furthermore, they internalize stakeholders' and citizens' voices through societal feedback to conduct research that gains social support (societal-level collaboration).

Participating in multiple levels of collaboration is crucial not only for advancing science as a whole but also for individual researchers to develop into scholars who can realize such collaboration through internal dialogue. In researcher development, particularly in today's world where science and society increasingly intersect, it is essential to provide experiences of multiple levels of collaboration, and to cultivate the mindset to research while navigating multiple perspectives gained from these experiences.

Intensified competition among researchers activates team-level identities directly linked to achievements. If one identity becomes dominant, it can lead to research misconduct due to obsession with personal achievements, or result in self-righteous, closed communities bound by specific interests (exemplified by Japan's "nuclear village"—a network of politicians, corporations, and researchers who mutually benefited from promoting nuclear power, which was accused of being responsible for the Fukushima nuclear power plant accident that occurred right after the Great East Japan Earthquake in 2011). Regularly examining one's research from multiple identity perspectives is valuable in determining one's direction.

Future Prospects: Co-creation of Society and Science

Transformational Creativity aims to have creativity generate a sustainable impact on communities and the environment beyond individual benefits (Sternberg, 2021). The emphasis is on promoting long-term social transformation rather than individual gains or short-term results. Responsible Research Innovation (RRI) originated in the EU and is gaining traction in Japan. This concept seeks to promote the transformation of research and innovation systems by aligning their development with societal values and needs. It emphasizes collaboration among social actors, including researchers, citizens, policymakers, businesses, and NGOs, throughout the entire research and innovation cycle. The approach prioritizes principles such as diversity and inclusiveness, foresight and reflexivity, responsiveness, and openness and transparency (Owen et al., 2012). Societal-level collaboration is essential for researching public interests and social challenges while aligning with societal values and needs.

As open science advances internationally, the interaction between science and society becomes increasingly dynamic. Consequently, team, community, and societal relationships are growing more complex and fluid. In the fatty acid analysis case from section "Collaboration at the Societal Level", what began as societal-level collaboration on a bulletin board gradually evolved into a new research team. Similarly, discussion communities based on bulletin boards can become the seeds of new communities from societal-level collaboration.

Society can generate multifaceted ideas and participate in their selection and refinement. For instance, nuclear power plant safety is evaluated not only by nuclear scientists and engineers but also by experts from various fields, and by residents. Regarding this, Fujigaki (2003, p. 102) presents the concept of public validity boundaries in contrast to scientists' validity boundaries, examining the gap between them. Society can also indirectly influence idea production, selection, and refinement by teams and communities, as exemplified by the manifestation of shared praxis and crowdfunding support discussed in section "Collaboration on Electronic Bulletin Boards: New Questions Emerging from Dialogue with Society". Since scientific activities are inherently part of societal activities, they are influenced by social context.

What role is science expected to play in dialogue with society? In this case study, the shared praxis of science in general was emphasized at the societal level. It suggests that science's social role and value lie in creating knowledge according to scientific principles without yielding to sponsors, even when confronting inconvenient truths. Society is not always right. For example, the availability cascade is a mechanism through which society's biases about risks become incorporated into public policy (Sunstein & Kuran, 2007). This process begins with media coverage of minor incidents and develops into public panic and government intervention. This phenomenon will likely cause even more serious situations in today's world, where social media is widespread. This research indicates that society as a whole must engage in healthy interactions with others who hold different and diverse opinions under the shared praxis of democracy.

How can trust in science be built while fulfilling this role? This case study has revealed deepening mutual understanding through interactions between archaeologists and citizens on a bulletin board. Archaeologists witnessed firsthand how research misconduct damages trust in their field and related disciplines, the severity of societal criticism, and the resulting confusion in educational and social settings. Citizens, in turn, gained insight into the conditions under which archaeologists conduct research and verification. Mutual understanding accompanied by concrete awareness of each other's existence and the impact of their actions proves effective in building trust.

The bulletin board also revealed confusion in various practical contexts, including education, cultural property administration, and excavation work. Emerging problems and concerns that could become significant issues in the future may be visible to only a few individuals. Distortions often become more apparent at the periphery rather than the center. The negative social impacts of research misconduct are sometimes more keenly felt by those working directly with society than by university researchers. Such latent societal needs can be discovered through bulletin board interactions.

In order to create knowledge that benefits the public in society, trust must be established on multiple levels: as experts, scientists, and citizens. This trust is formed through team, community, and societal collaboration. A perspective encompassing multiple levels of collaboration is valuable in creating environments and systems that generate positive creativity and innovation for society and in constructing individual researchers' identities. We hope this chapter contributes, even modestly, to realizing the vision of science bringing about positive social change.

References

Andersson, J., & Rönnberg, J. (1996). Collaboration and memory: Effects of dyadic retrieval on different memory tasks. *Applied Cognitive Psychology, 10*, 171–181. https://doi.org/10.1002/(SICI)1099-0720(199604)10:2<171::AID-ACP385>3.0.CO;2-D

Baldwin, M. (2015). Credibility, peer review, and Nature, 1945–1990. *Notes and Records, 69*(3), 337–352. https://doi.org/10.1098/rsnr.2015.0029

Basden, B. H., Basden, D. R., Bryner, S., & Thomas, R. L. (1997). A comparison of group and individual remembering: Does collaboration disrupt retrieval strategies? *Journal of Experimental Psychology: Learning, Memory, and Cognition, 23*, 1176–1189. https://doi.org/10.1037/0278-7393.23.5.1176

Campbell, D. T. (1960). Blind variation and selective retention in creative thought as in other knowledge processes. *Psychological Review, 67*(6), 380–400. https://doi.org/10.1037/h0040373

Chevalier, A., Bonnardel, N., & Piolat, A. (2000). Interactions on the Web between reviewers and authors: Towards reviewing an article for the e-journal "JIME". In *Proceedings of the 10th annual meeting of the society for text and discourse* (pp. 128–129).

Clark, C. J., Costello, T., Mitchell, G., & Tetlock, P. E. (2022). Keep your enemies close: Adversarial collaborations will improve behavioral science. *Journal of Applied Research in Memory and Cognition, 11*(1), 1–18. https://doi.org/10.1037/mac0000004

Cowley, S. J. (2015). How peer-review constrains cognition: On the frontline in the knowledge sector. *Frontiers in Psychology, 6*, 1706. https://doi.org/10.3389/fpsyg.2015.01706

Csikszentmihalyi, M. (2014). The systems model of creativity and its applications. In D. K. Simonton (Ed.), *The Wiley handbook of genius* (pp. 533–545). Wiley Blackwell. https://doi.org/10.1002/9781118367377.ch25

Curhan, J. R., Labuzova, T., & Mehta, A. (2021). Cooperative criticism: When criticism enhances creativity in brainstorming and negotiation. *Organization Science, 32*(5), 1256–1272. https://doi.org/10.1287/orsc.2020.1420

Cyranoski, D. (2000). Fake finds reveal critical deficiency. *Nature, 408*, 280. https://doi.org/10.1038/35042725

Dunbar, K. (1993). Concept discovery in a scientific domain. *Cognitive Science, 17*(3), 397–434. https://doi.org/10.1207/s15516709cog1703_3

Dunbar, K. (1997). How scientists think: Online creativity and conceptual change in science. In T. B. Ward, S. M. Smith, & S. Vaid (Eds.), *Conceptual structures and processes: Emergence, discovery and Change* (pp. 461–494). APA Press.

Dunbar, K., & Schunn, C. D. (1990). The temporal nature of scientific discovery: The role of priming and analogy. In *Proceedings of the twelfth annual meeting of the cognitive science society*. Erlbaum.

Fujigaki, Y. (1998). Filling the gap between the discussion on science and scientists' everyday activities: Applying the autopoiesis system theory to scientific knowledge. *Social Science Information, 37*(1), 5–22. https://doi.org/10.1177/053901898037001001

Fujigaki, Y. (2003). *Senmonchi to kōkyōsei [The public ethics and the spirit of specialism]*. University of Tokyo Press. (In Japanese).

Hermans, H. J. M., Kempen, H. J. G., & Van Loon, R. J. P. (1992). The dialogical self: Beyond individualism and rationalism. *American Psychologist, 47*(1), 23–33. https://doi.org/10.1037/0003-066X.47.1.23

Homma, M. (1996). Reconsidering group productivity in brainstorming. *Japanese Psychological Review, 39*(2), 252–272. https://doi.org/10.24602/sjpr.39.2_252. (In Japanese).

Iseda, T. (2001). Scientific rationality and two concepts of 'social'. *Studies in Informatics and Sciences, 14*, 27–42. (In Japanese).

Japanese Archaeological Association. (2003). *Inspection of the early and middle Palaeolithic problem in Japan (Zen, Chuki Kyusekki Mondai no Kensho)*. Japanese Archaeological Association. (In Japanese).

Kelly, J., Sadeghieh, T., & Adeli, K. (2014). Peer review in scientific publications: Benefits, critiques, & a survival guide. *EJIFCC, 25*(3), 227–243. https://www.ncbi.nlm.nih.gov/pmc/articles/PMC4975196/

Kovach, B., & Rosenstiel, T. (2014). *The elements of journalism: What newspeople should know and the public should expect* (Rev. and updated 3rd ed.). Three Rivers Press.

Latour, B. (1987). *Science in action: How to follow scientists and engineers through society* (K. Kawasaki & K. Takada, Trans.). Sangyo Tosho, 1999. (Original work published 1987).

Mainichi Shimbun. (2000a, November 5). *Mainichi Shimbun Morning Edition*, (In Japanese).

Mainichi Shimbun. (2000b, November 15). *Mainichi Shimbun Morning Edition*, (In Japanese).

Marion, S. B., & Thorley, C. (2016). A meta-analytic review of collaborative inhibition and postcollaborative memory: Testing the predictions of the retrieval strategy disruption hypothesis. *Psychological Bulletin, 142*(11), 1141–1164. https://doi.org/10.1037/bul0000071

Mullen, B., Johnson, C., & Salas, E. (1991). Productivity loss in brainstorming groups: A meta-analytic interaction. *Basic and Applied Social Psychology, 12*(1), 3–23. https://doi.org/10.1207/s15324834basp1201_1

Nicholas, D. (2015). Peer review: Still king in the digital age. *Learned Publishing, 28*(1), 25–29. https://doi.org/10.1087/20150104

Nielsen, M. (2013). *Reinventing discovery: The new era of networked science* (Y. Takahashi, Trans.). Kinokuniya Shoten, 2013. (Original work published 2011).

Oda, S., & Keally, C. T. (1986). A critical look at the Palaeolithic and "Lower Palaeolithic" research in Miyagi Prefecture, Japan. *Journal of the Anthropological Society of Nippon, 94*(3), 325–361. https://doi.org/10.1537/ase1911.94.325

Okada, T. (1999). Collaborative research processes in science: An examination through interview, questionnaire, and psychological experiment. In T. Okada, H. Tamura, K. Todayama, & K. Miwa (Eds.), *Thinking about science: A Multidisciplinary perspective* (pp. 2–25). Kitaohji Shobo. (In Japanese).

Okada, T., & Simon, H. A. (1997). Collaborative discovery in a scientific domain. *Cognitive Science, 21*(2), 109–146.

Osborn, A. F. (1957). *Applied imagination: Principles and procedures of creative thinking* (2nd ed.). Scribners.

Owen, R., Macnaghten, P., & Stilgoe, J. (2012). Responsible research and innovation: From science in society to science for society, with society. *Science and Public Policy, 39*(6), 751–760. https://doi.org/10.1093/scipol/scs093

Ozawa, C. P., & Susskind, L. (1985). Mediating science-intensive policy disputes. *Journal of Policy Analysis and Management, 5*(1), 23–39. https://doi.org/10.2307/3323411

Sato, S. (2016). Peer review: Issues and antidotes. *The Journal of Information Science and Technology Association, 66*(3), 115–121. https://doi.org/10.18919/jkg.66.3_115. (In Japanese).

Schunn, C. D., Crowley, K., & Okada, T. (1998). The growth of multidisciplinarity in the Cognitive Science Society. *Cognitive Science, 22*(1), 107–130. https://doi.org/10.1016/S0364-0213(99)80036-6

Sternberg, R. J. (2021). Transformational creativity: The link between creativity, wisdom, and the solution of global problems. *Philosophies, 6*(3), 75. https://doi.org/10.3390/philosophies6030075

Sumner, T., Buckingham Shum, S., Wright, M., Bonnardel, N., Piolat, A., & Chevalier, A. (2000). Redesigning the peer review process: A developmental theory-in-action. In R. Dieng, A. Giboin, G. De Michelis, & L. Karsenty (Eds.), *Designing cooperative systems: The use of theories and models* (pp. 19–34). I.O.S. Press.

Sunstein, C. R., & Kuran, T. (2007). Availability cascades and risk regulation. *John M. Olin Program in Law and Economics Working Paper*, 364. https://chicagounbound.uchicago.edu/law_and_economics/70/

Thagard, P. (1997). Collaborative knowledge. *Noûs, 31*(2), 242–261. https://doi.org/10.1111/0029-4624.00044

Todayama, K. (2020). *Kyōyō no sho [The book of enlightenment]*. Chikuma Shobō.

Ware, M. (2017). What might peer review look like in 2030? *Figshare*, 2. https://doi.org/10.6084/m9.figshare.4884878

Weldon, M. S., & Bellinger, K. D. (1997). Collective memory: Collaborative and individual processes in remembering. *Journal of Experimental Psychology: Learning, Memory, and Cognition, 23*(5), 1160–1175. https://doi.org/10.1037/0278-7393.23.5.1160

Wiley. (n.d.). *Types of peer review*. Wiley Author Services. Retrieved January 23, 2025, from https://authorservices.wiley.com/Reviewers/journal-reviewers/what-is-peer-review/types-of-peer-review.html

World Wide Views. (n.d.). *World wide views*. Retrieved January 23, 2025, from https://wwviews.org/

Yamanouchi, Y. (2017). The roles of open scientific communication for research integrity. *Journal of Science and Technology Studies, 14*, 63–76. https://doi.org/10.24646/jnlsts.14.0_63. (In Japanese).

Yamanouchi, Y., & Okada, T. (2003a). Construction of a validity boundary in an academic community: A new direction in studies of scientific thinking. *Cognitive Studies, 10*(3), 418–435. https://doi.org/10.11225/jcss.10.418. (In Japanese).

Yamanouchi, Y., & Okada, T. (2003b). Formation process of validity-boundary: A case study in an archeology Community. *Journal of Science and Technology Studies, 2*(2), 23–41. (In Japanese).

Yamanouchi, Y., & Okada, T. (2003c). Dynamics of formation of validity-boundary: Focusing on a social factor. *Journal of Science and Technology Studies, 2*(2), 42–56. (In Japanese).

CHAPTER 6

The Story Lab: What Can We Learn from Embodied Cognition?

Samira Bourgeois-Bougrine and Antoine Le Bos

Introduction

Stories typically recount a sequence of real or fictional events in which one or more protagonists interact with their world, often confronting and attempting to resolve problems along the way (Box 6.1). Storytelling, which represents the social and cultural activity of sharing stories, is likely the oldest form of teaching, allowing generations of humans to share cultural knowledge to be remembered over time (Landrum et al., 2019). Indeed, the functions of stories have been recognized for centuries. For instance, anthropological work on hunter-gatherer communities links storytelling to enhanced survival outcomes, reproductive success, and group cooperation through sharing warnings of danger, explaining routes to resources, exchanging social information, etc. (Lee et al., 2020). Several

S. Bourgeois-Bougrine (✉)
Université Paris Cité and Univ Gustave Eiffel, LaPEA, Boulogne-Billancourt, France
e-mail: sbb9799@gmail.com

A. Le Bos
Le Groupe Ouest, Plounéour-Brignogan, France
e-mail: antoine.lebos@legroupeouest.com

N. Bonnardel (ed.), *Collaborations in Creativity*, The Seven C's of Creativity, https://doi.org/10.1007/978-3-032-07753-0_6

Box 6.1 What Qualifies as a Story?

In this project, we define a **story** as the verbal or nonverbal discourse of a sequence of inter-related events, either fictional or nonfictional (e.g., creation of the imagination or as a truthful account of facts).

Narrative derives from the Latin *narrare*, meaning "telling," and gnarus, which means "being acquainted with something," "having knowledge of something" (Kvernbekk, & Frimannsson, 2013). Narrative implies knowledge (knowing), gained through life experience. Derived from it is the word narratio, which means story.

Narrative can be divided into two aspects (Chatman, 1978): (1) **the story itself, the content**, and (2) **the manifestation, the expression, or the way it is told** (play, short novel, movie, cartoon, epic, tales, etc.). Simply put, it is a distinction between the things told and the ways of telling them.

Key elements of stories: plot, characters, conflict, message. The **plot** is the chain or sequence of events that affect each other through the principles of cause-and-effect. The plot line is supported by the characters' intentions and motivation, etc. The **characters** are the driving force in fiction. The conflict is about opposition and struggle between characters with opposing goals and desires, a character versus nature, society, its own doubts, contradictions, etc. The **message**, which is a central moral or ideological statement, is gained through the resolution of the conflict. It is where you communicate the perception of right and wrong.

Storytelling is the social and cultural activity of sharing stories and it refers generally to oral storytelling.

studies have explored how sharing knowledge through narratives could affect memory, brain, attitudes, beliefs, and behaviors. Listening to or reading a story tends to have more depth than a simple example in the sense that it engages our thinking, our emotions, and can even lead to the creation of mental imagery (Green & Brock, 2000). For instance, a recent study that explored the neurosciences of how we transmit memories of a movie to others showed that the listeners' brain "lights up" exactly as the storyteller's brain (Zadbood et al., 2017). This means that a recalled narrative of a movie triggered a mental construction of the events (e.g., imagination) in the listeners' brain and enabled them to partly experience a movie they had never watched.

Where those stories came from and how they are developed? The fascination of the creative writer as an individual genius led to a considerable amount of research in psychology dedicated to the study of writers' personality, individual mental processus, etc. According to Sawyer (2006, p. 206), "writing seems to be far removed from social and contextual influences. You do not need anyone's help to write poetry; you do not need to use complex tools or to collaborate in a system of cooperative work." Although the myth of a story authored by a lone genius is relatively dominant, research suggests that the generation and development of vivid and engaging stories is fundamentally a collaborative creative act. Indeed, researchers are increasingly acknowledging that stories emerge through a creative collaboration where cognition is distributed. A popular example of creative collaboration is the famous Oxford University "Inklings."[1] Among its ranks, The Inklings included authors such as J.R.R. Tolkien, author of "The Lord of the Rings," or Lewis, well known for his book "Out of the Silent Planet." This informal literary discussion group, "*provided a supportive and stimulating environment for its members, fostering creativity and collaboration…each member contributed their unique insights and perspectives to help refine and enrich their peers' works, too.*"[1]

Moreover, creative writing has been mostly considered from the classical cognitive approaches to mind in terms of mental representations and internal computations. Mental representations are information-containing, physically realized structures in the brain; existing mental representations influence how we perceive, make sense of, process, understand, and recall all stimuli (Krcmar & Haberkorn, 2020). According to Wilson and Golonka (2013, p. 2), "*the standard cognitive psychology explanation for the form of* [an observed] *behavior is that it reflects the contents and operation of an internal algorithm (implemented as a mental representation) designed to produce that behavior on demand.*" Predictive processing which is one of the most prominent theories in cognitive and computational neuroscience postulates that, because the mind "lacks access to the external world," its task is to constantly generate and update a "mental model" or representation of the environment through the prediction of the properties of the sensory input. If the prediction comes close to or matches the

[1] The Inklings: Celebrating Oxford's Fantasy Literature Trailblazers.

sensory input, the brain has a good grasp on the world (Gallagher, 2023). According to Kirchhoff (2018, p. 2357), "*predictive processing states that organisms must act so as to maintain themselves within expected states, and that organisms can do so only by minimizing a prediction error quantity, while, at the same time, seeking to estimate and make more precise the reliability of the prediction error signal itself.*" Therefore, the overarching principle of the theory of predictive processing is the minimization of surprise or, equivalently, the maximization of expectation. Clark (2018) considers that the best way to achieve it is to live in a boring, predictable world such as a Dark Room. He challenges the claim that considers human brains as devices that strive to minimize prediction error between actual and predicted sensory stimulations. According to him, this predictive framework is in contrast with human "deep positive attractions of novelty, play, and exploration." He emphasizes the role of the "epistemic foraging, and cultural practices in shaping the restless, curious, novelty-seeking human mind."

In contrast with the traditional theories of cognition that emphasize the body as "*a 'passive' observer to the brain, and necessary only in the execution of motor actions*" (Fugate et al., 2019, p. 1), embodied cognitive (EC) sciences consider that human cognition is not solely in the head and that the states of the body modify the states of the mind (Wilson & Golonka, 2013). EC is a result of dynamic interactions between the brain, the body (head and body movements, postures, etc.), and all the dynamic interactions with the social environment (e.g., presence of an audience) as well as the physical world itself. Embodiment theories emphasize the body functioning as a "*constituent of the mind*" rather than secondary to it (Fugate et al., 2019). EC gives "*primacy to embodied activity over and above the re-construction of an outside world on the basis of internal, mental representations*" (Kirchhoff, 2018, p. 2356). Referring to dancers, Satama et al. (2022, p. 168) define creative collaboration as "*a collective social phenomenon that emerges from the constant negotiation of meanings, ideas and interpretations of one another in bodily ways, that is, in ways that highlight sensuous, empathetic and reflexive ways of being in the world in relation to others.*" Similarly, we consider collaborative creative writing as a complex, messy, situated, and embodied process in which ideas arise by listening, sensing, empathising with others and interacting with artifacts and the environment.

In this chapter, we will start with a brief introduction to the "4E" perspective of embodied cognition (embodied, embedded, extended, and enactive). Then, we will attempt to shed light on the initial phases of stories development from both perspectives: staged information processing in the case of "solo" scriptwriting for films and embodied cognition experiences more adapted to TV series codevelopment where several writers cowrite at the same time. More specifically we will discuss this type of creative collaboration based on the "4E" perspective. By doing so, we will provide academic insights as well as a glimpse into the embodied experiences developed recently by Le Groupe Ouest (LGO). Among LGO's recent experiments, we will specifically take a close look at a new set up for writers' collaboration in ideation phase called "the GREENHOUSE" within the story elaboration phase. LGO's experiences involved the conception of a physical narrative space where professional writers can enter, explore, experiment, and experience all the possibilities, at the same time. The concept of the GREENHOUSE is the opposite of the Dark-Room scenarios (Clark, 2018; Friston et al., 2012). As we will present in this chapter, writers immersed themselves in the GREENHOUSE narrative space because it engineered the unexpected, created surprises and new meaning, and allowed the writers to "walk through" their stories in specialized narrative 3D space. Different realities of embodied collaboration will be discussed and suggestions of possibilities for future research will be provided.

Introduction to Embodied Cognition

Embodied cognition (EC) has its roots in various domains such as neurosciences behavioral study, linguistics, philosophy, psychology, biology, developmental and evolutionary sciences, and robotics. This diversity creates richness but generates complicated theories and concepts. EC research increasingly considers several facets of cognition such as Embodiment, Embedded, Extended, Enactive, Ecological, Emotional, Empathic, and Existential. However, among the many theories of EC, we will focus in this chapter on the most influential concept of 4E (Embodiment, Embedded, Extended, and Enactive). To paraphrase Slaby (2014), EC is controversial enough to talk about more counterintuitive Es.

The 4E considered in this chapter have many convergences such as the centrality of body-environment coupling but defined in different ways. Among those convergences, the prominent ones are the similarities between extended and enactive approaches. In addition, there are differences within each of the Es such as weak and strong Embodiment or the three versions or waves of theories related to extended mind which further complicate the concept of EC, etc. (see Gallagher, 2023, for deep analysis of the convergences and limitations of the 4E concept).

The first E, "Embodiment," refers to how the body anatomy, movements, and somaesthetic factors contribute to cognitive processes. Weak embodiment considers the brain or mental representations in bodily formats (or B-formatted representations) rather than the body anatomy or its movements that are involved in the cognitive process (Goldman & de Vignemont, 2009). In contrast to weak embodiment, the claim of strong embodiment is that the way our heads and bodies move shape our cognitive experience prior and after the brain processing. For instance, enacting embodied metaphors such as "think out of the box" (Leung et al., 2012) or "give your ideas some legs" (Oppezzo & Schwartz, 2014) improves creativity.

The second E, "Embedded," introduces the concept of affordances (e.g., opportunities for action) and covers the idea of how the environment scaffolds our cognitive processes, and how the manipulation of the environment can reduce our cognitive load. Embedded cognition includes different approaches such as situated cognition, distributed cognition, and ecological cognition. The main idea is that individuals are actively or passively situated in the environment which can enable or restrain cognition. Moreover, "situated" individuals could engage actively in actions to scaffold cognition or reduce cognitive load.

The third E, Extended, considers how the material we use not only enables cognition but transforms the very nature of cognition (Gallagher, 2023). One of the main perspectives defended by Extended mind theorists is integration. From this perspective, the use of notebooks, smart phones, writing tablets, diagrams, maps involves more than simple offloading or scaffolding that simplifies cognition (as considered in embedded cognition). Such engagements induce a transformation of our cognitive processes such that we can think and act in new ways.

The fourth E, "Enactive," emphasizes the idea that perception is for action and that this action orientation shapes most cognitive processes (Malinin, 2019). To paraphrase Gallagher (2023) statement, Enactivists endorses non-representational approaches; perception and action are treated as basic cognitive processes that scale up to more complex forms of sense-making via intersubjective (social and cultural) practices. They take intersubjective interactions to be fully embodied perceptual and motoric processes, involving facial expression, posture, movement, gestures, vocal intonations, and specific forms of sensory-motor couplings in highly contextualized pragmatic and social environment.

Creative Writing: From Occasional Collaboration to Cocreation

The scenario as a creative object implies setting in motion and generating in oneself, worlds, spaces, bodies, beings and their sensibilities, conflicts, movement. However, our imagination is at times limited by personal and contextual factors which reinforce the need to use the most "activating" methods possible. This section will present how embodied social interactions, feedback, collaboration, and cocreation enhance the ability to generate multiple ideas, alternatives, possibilities, and exploring problems from various perspectives.

Script "Solo" Writing, Social Feedback, and Interaction

Creativity-in-the-wild or real-world creativity *is "a situated practice; it involves embodied experiences and is embedded in socio-material environments. Dynamical systems approach also describe how creativity is enacted through person-environment interactions and distributed among technological-material artifacts and other people*" (Malinin, 2019, p. 9). However, for decades, the reigning paradigm of cognitive science has been the computational cognition (CC) model of the mind which describes cognition as a staged information processing: perception, then cognition, and finally action (Malinin, 2019; Goldman & de Vignemont, 2009). The CC model has been nicely captured by Hurley "classical sandwich model" of the mind where the "outer slices of perception and action are peripheral to the inner filling of cognition, and thus separate from one another. They are also separate from cognition, which interfaces between perception and action" (Burr, 2017, p. 3).

In our study of the creative process of screenplay writing (Bourgeois-Bougrine et al., 2014), we partly constructed our interview guide on to cover scriptwriters' activity in terms of information input (where and how the scriptwriter gets information and inspiration), mental processes (reasoning, planning, thinking, daydreaming, problem-solving, decision-making), and the output of those mental processes. The main stages of scriptwriting, as they appear from our interviews, and the psychological processes screenplay writers refer to in each one of them are outlined in Fig. 6.1. Two observations are required at this point. First, these phases are not seen as separate from each other but interrelated, with the possibility of returning to a previous phase, jumping ahead or engaging in more than one phase at a given time. Second, it is to be noted that we are not operating here with a strict separation between the psychological and the social. On the contrary, we conceive of these as inter-related and coevolving facets of any human activity system (Cole, 1996) since social interactions and environments are both shaped by and actively shaping the psychological makeup and responses of the person, even in moments when the person is thinking or working alone (see Marková, 2003). Although

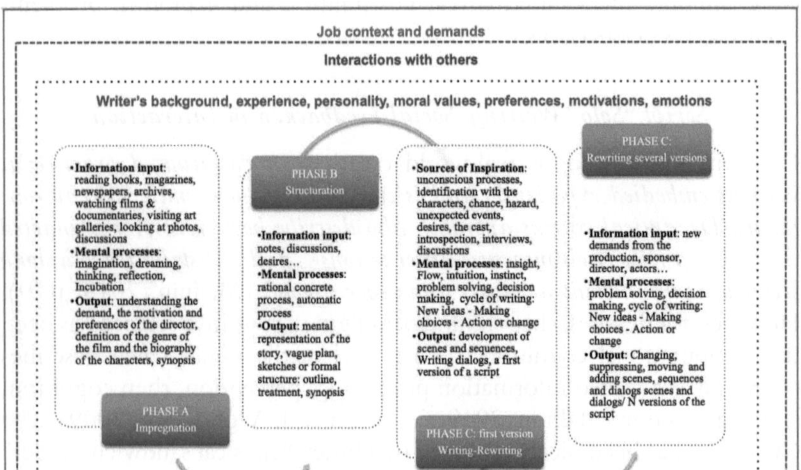

Fig. 6.1 Activities and stages of the writing as well as interactions with others, job context, and demands

we considered the impact of the relationships with others (director, producer, sponsors, producer, actors, film crew, family and friends, etc.) and the job context and constraints on the creative process, we positioned an adapted version of the "sandwich model of mind" at the heart of creative writing model (in blue in Fig. 6.1). Each of the three phases of the creative process (A, B, and C) starts with information input or inspiration sources (e.g., perception), then the mental processes (e.g. cognition), and finally the output or actions.

William Goldman mentioned that "the writing is never the problem, it is knowing what the structure is and what goes where and who the people are and how this scene relates to this… I have to know all that before I start" (Goldman, 1996 pp. 12–13). At the beginning of private writing episodes, participants commented on their difficulty to get started (Bourgeois-Bougrine et al., 2014, p. 7):

- "I am starting a new project and currently I am turning around. It's a kind of ritual: I am turning around the topic like a wolf around its prey."
- "I find it hard to tackle the job head on. I first circle, I turn around. I have a relation to work a bit like the way I read the newspaper: when I spot a page in a newspaper that interests me, I first read all those with little interest to me to finish by reading the one that interests me. This is a very perverse mania; when I find a good idea, or a scene that I like, I turn around."

To quote Getzels (1979, p. 167), ideas do not present themselves as "problems capable of resolution or even sensible contemplation. They must be posed and formulated in fruitful and often radical ways if they are to be moved toward solution. The way the problem is posed is the way the dilemma will be resolved." Einstein (in Getzels, 1979) put it: "The formulation of a problem is often more essential than its solution." Although our model is valid mostly for "solo" writing for films, we will explore in the next section an embodied collaborative experience, developed by Le Groupe Ouest (LGO) that is more in line with the process of cowriting TV series.

Script Codevelopment

Stories are most often built individually, but in LGO, collaborative processes are activated to reinforce the strength and substance of the narrative. Within the 20-year practice of hundreds of screenwriters developing their stories in residency in Le Groupe Ouest, it has become obvious that writers benefit enormously from confronting the vision of their story shapes or paths with their peers' understanding and projections. To illustrate the concept of the 4E of embodied cognition, we will delve into the embodied experiences developed by Le Groupe Ouest (LGO) that evolved throughout the years from collaborative workshops to the conception of a physical narrative space where professional writers can enter, explore, experiment, and experience all the possibilities, at the same time.

Collaborative Workshops

It all started in LGO, on the coast of Northern Brittany—Europe's leading coaching center for screenwriters and filmmakers at writing stage. Over two thousand screenwriters from all over the world have been coached through in-residency workshops in 20 years. Around 2010, already more than a hundred screenwriters and filmmakers had been coached at LGO. Each workshop group was built around eight projects, some of these projects being cowritten. At that time, all the projects entered the game with a written draft which consisted most of the time of a "treatment" (summary of actions) of five to ten pages. The workshops included seven to ten fellow writers or filmmakers around the table in addition to two coaches. Each time a writer dives inside a collective process, more and more often taking the shape of a writer's room around each project during daily slots of 45–90', the result in terms of opening of enriched visions and possibilities is systematically breathtaking. Each pair of coaches was chosen for the complementarity of the metaphoric and analytic tools they brought to the group. Often a coach (an experienced professional screenwriter) is chosen to bring a singular field of analysis developed through their upbringing or graduate studies (physics, mathematics, biology, architecture, etc.), the other coach is chosen to bring another field heterogeneous to the previous (psychoanalysis, psychology, theater, visual arts, etc.). The core methodology of coaching was based around circular discussions and feedback. To enact the metaphor "give your ideas some legs" (Oppezzo & Schwartz, 2014), one of LGO key methods to unlock possibilities is to brainstorm with other group members while walking on the nearby limitless beeches potentially for hours without obstacles. In addition, at the end of each day of the workshop, the two LGO coaches spend an hour debriefing with the core team, in order to understand what kind of progress or resistance the writers and the projects generated.

In 2010, it became clear for the core pedagogical team of LGO that the use of the actual act of writing in the idea development phase is often a hindrance to the most creative possibilities, as well as a hindrance to collaboration. As if Gutenberg wasn't as much of an ally to restless minds as we think. The written track of the project, which should have been considered a mere hypothesis for the writer, was at the core of a strong resistance phenomenon to any rethinking process. As if because an idea is written, it is carved in stone. Which in a way is in the brain, because of the written track! This phenomenon could be related to the psychological concept of fixation where the generation of alternative creative ideas could be limited by recently activated knowledge (Wang et al., 2023). These counterproductive effects of prior knowledge could be enhanced when it involves affect. For instance, one scriptwriter mentioned in our research (Bourgeois-Bougrine et al., 2014, p. 12) that a part of the reason of fixedness is "*because in the script, there is already the affect: we started to love our scenes, to make our characters speak; we become attached to a particular moment, a particular dialogue...we then struggle to understand where the problem is or to cut when it is too long.*" In the process of cowriting and sharing knowledge, each party needs to understand the other's point of view while taking distance from one's own as a precondition for any successful collaboration (Bourgeois-Bougrine & Glăveanu, 2018). As another script writer put it: "*One should not be in the projection of the self in a scene so that, if one amends it, it's like a personal prejudice. Yet, many people are like that*" (Bourgeois-Bougrine et al., 2014, p. 12).

To skip over the drawback of the written tool, LGO team decided to make extensive use of orality as a propellant of creative possibilities. As part of its European Think Tank (StoryTANK), the team discovered through the research of Bernard Victori (research director in linguistics at the CNRS) that orality was propelled by the need to tell stories. This means that anyone who masters an oral language masters a free software not only to tell stories, but also to invent them. A godsend for scriptwriters, but also for all humans. However, orality introduces blind spots: what does this character look like? How is this village laid out? What hidden possibilities and points of view lie behind this first idea? LGO's pedagogical team decided to test the use of huge boards where writers could draw possibilities, improvise collages of printed faces and situations, map places and pieces of worlds and human trajectories. Screenwriters were a bit afraid at first by this metamorphosis of their toolbox, but most of them quickly ended up installing big paper boards at home, as well as installing mapping boards in their offices, as if they were becoming detectives inside their own projects, starting to see themselves as doing an enquiry inside their story world and story possibilities, looking for the most beautiful and intense story option.

Orality was there to share the deployment of the projects externalized on the boards. But as soon as we had to work with screenwriters on Series projects came in the episodic logic and its complexity of multiple interacting trajectories. How to represent all these layers—temporal, spatial, human, aesthetic, narrative, and dramaturgical? It started with the addition of tracing paper on the boards, but it was not enough. The thickness of the serial material was cramped in there. We decided to launch an experience: letting series writers in writers' rooms use whatever surface they need in a room. The walls of course, but possibly the windows, the door, every possible surface. And the result came quickly: the teams of writers needed to spread their imaginary world into every corner of the room! Debrief after debrief, the obvious had to be taken into account: why don't we give teams of writers three-dimensional setups to spread ideas. This is how ***the GREENHOUSE*** was born!

Embodied Experiences for Screenwriters and Film Makers

The GREENHOUSE is a three-dimensional space that could be covered with a black blanket, plunging it into dim light or even in darkness. It includes three interconnected narrative spaces (prep room, divergence room, and convergence room) dedicated to boosting imagination at different stages of the creative process. For instance, the divergence room allows writers to populate the whole 3D space with artifacts, pictures, keywords, verbs, obsessions, themes, maps, collages, faces, drawings, sketches where the writer can enter, explore, experience, create meaning, etc. Some of those visual artifacts are generated live in prep room by artificial intelligence based on the writers "externalized" keywords or prompts (the Wizard of Oz-like generation). Moreover, the physical 3D space of the GREENHOUSE allows authors to "enter" their story and experience it from within. The images can be uploaded in VR headset to allow the writers to enter and be completely immersed in their story. By moving around the GREENHOUSE rooms and arranging elements, authors create a sense of both space and time that's difficult to achieve when writing on paper. This spatialization helps them visualize the "momentum" of the story and experience it from within. The 3D narrative space of the "GREENHOUSE" (1) supports spatial and embodied cognition, (2) supports authentic dialogue and creative collaboration (between different stakeholders from various backgrounds, based on oral narration), and (3) helps populating the space with rich and diverse material, helping "engineer" serendipity.

The embodied experience started in the prep room when the coaches asked trios of authors to agree on an initial idea of a story, a kind of a little seed of idea. This seed could be an image, text snippet, a personal memory, a dream, etc., but nothing too elaborate. Then, the coaches guided the writers to help them "externalize" keywords evoked by their initial idea, taking internal ideas and projecting them outward in a tangible form. In this early process of externalization, when a writer talks about the world they have in mind, a network of keywords is generated as a rhizome[2] that spreads without hierarchy and can be entered at any point. As one of LGO coaches put it, "*externalising things allows the author to create a distance between the object and himself. This distance creates a possibility of looking, of scrutinizing the object in completely different ways than when it's in there* (gesture toward the head). *But also, as soon as it's externalized, it allows it to be shared, to collaborate with others, it becomes a matter of common space. In addition, when a writer talks about the world they have in mind, fellow writers draw on a paperboard the world they 'see' from what they hear. The technical quality of the drawing is not the point.*"

The first exploration phase involved populating the GREENHOUSE to enrich the world and possibilities for the writers (see Fig. 6.2). All

Fig. 6.2 Populating the GREENHOUSE with images, keywords, drawings, etc.

[2] The concept of rhizome, developed by the philosopher Gilles Deleuze and the psychoanalyst Félix Guattari, refers to a structure that is constantly evolving, in all horizontal directions, and devoid of levels. This structure is opposed to the pyramid-shaped hierarchy (or "arborescence").

around the roof and the sides of the GREENHOUSE, the writers hanged pictures, keywords, verbs, obsessions, themes, maps, collages, faces, drawings, sketches, etc. To this end, the writers were given three inspirational sources of information to choose from to "populate the GREENHOUSE." The first set was an LGO collection of images of faces. The second source was a random selection of magazines from which pieces of articles, images, titles, and slogans will be extracted and cut out by the authors. The authors were invited to use whatever could be triggering possibilities to develop their idea. The third source of information was a set of artificial intelligence images generated live by a visual artist who listened remotely to the writers' keywords. The rationale behind the AI-generated images was to "engineer" serendipity or the unexpected that will serve creativity and the generation of ideas. To paraphrase one of LGO coaches: *AI images are meant as creativity tools, not end products. The gaps between what authors imagine and what the AI produces can inspire new directions. And in fact, in these gaps, there is possible creativity.*

The writers were invited to populate the GREENHOUSE in a spontaneous and intuitive way, by artifacts, people, places, shapes, obsessions, dreams, etc., without overthinking. Reflecting afterward on his experience, one writer said: *When I started hanging those wasps, they were just needed there. Maybe also to make it more difficult to move around* [hanged red strings with clothes pegs at the end to look like flying wasps (see Figs. 6.3 and 6.5)]. This a typical example of what specialists of embodied cognition call "doing without representation" where the action (hanging the wasps) was not initially dictated by a mental representation or some internal beliefs but only through action. Like improvisation in music or theater, populating the GREENHOUSE takes the form of a continuous a cycle between action, perception, and cognition.

The second exploration phase invites the writers to consider physically the story from different perspectives, by observing from various angles, they are encouraged to freely associate ideas, rearrange images, and build on each other's contributions (Fig. 6.3). As one of the writers put it: "*When you have a board you see what everybody puts on but now, I'm doing something different: I am looking everywhere, from different angles, noticing something new, asking myself, or others, questions: Okay, what is that? Oh, that's in a good space, good spot…*" This phase involves covering the GREENHOUSE with a black blanket, plunging it into dim light or even darkness. And there a new game starts, with a flashlight (Fig. 6.4) for each writer, one after another. The writers were encouraged trying to find

Fig. 6.3 Exploration of the GREENHOUSE from different perspectives

Fig. 6.4 The flash-light phase within the "GREENHOUSE"

paths, chains of causes and effects through moving inside that space with the body of a character, as well as the body of the one doing the inquiry. One can use strings of wool to link key words and pictures, tying characters with desires, spaces, emotions, etc. To paraphrase Gaston Bachelard (1994), the GREENHOUSE offers a dwelling-place or 3D space that "shelters daydreaming,...protects the dreamer, ...allows one to dream in peace."

It quickly became obvious that this immersion within "the GREENHOUSE" became an accelerator of both creative and collaborative processes. The little seed starts to generate lots of paths, lots of possibilities. It activates a collaborative dynamic around it and it also makes the authors immediately start to be in a body dynamic. For instance, one of the trios started developing a character—a disruptive "wild man" who brings an unpleasant smell to a refined "closed" setting. They walk in their idea, they circulate in the idea, they connect, they create a fission, and by doing so, of course, a lot of possibilities arise as you can read in the exchange between a trio of writers in the GREENHOUSE:

- Writer 1: *There's something... You have the smell of nature, so he kind of brings this acid decomposing, a vinegary composting apple smell.*
- *Writer 2: And people are kind of sniffing their wine glasses like thinking it's that but it's him.*
- *Writer 1: And yeah, he kind of disrupts this façade* [gesturing to a large sheet hanging in the center of the GREENHOUSE where they crafted previously sophisticated crowds in a theater space]. *Notice this smell that he's brought with him and everyone's kind of checking their wine glasses because it smells like vinegar.*
- Writer 3: *Yeah. I am making wasps. / They do look a bit like wasps. They are flying around this man, so he is definitely not supposed to be there. That's the main thing. This man, huh? This man, yeah. He's not...Somebody should remove this man.*

The GREENHOUSE main functions are to create surprises and meanings, and the apparent chaos in the GREENHOUSE is fundamentally functional. Similar to oral preliterate people, poetry was not "produced primarily for enjoyment or to be aesthetically pleasing but is above all functional: indeed, it is the only means of storing, sharing and prolonging knowledge...Homeric Greeks used poetry as a kind of encyclopaedic databank: 'the whole memory of a people was poeticised, and this exercised a constant control over the ways in which they expressed themselves in casual speech" (Lloyd, 2023, p. 308). Within the GREENHOUSE, a

Fig. 6.5 "Entering" the story in the 3D narrative space

fictional space is generated, like a theater piece unfolding in front of the writers, triggered by the fact that they stepped from their inner spaces (brains) into a proper three-dimensional space of exploration. Under dim light, in the dense and populated space of the GREENHOUSE, surprising encounters or discoveries are made, new possibilities and meanings emerge, life appears, thanks to the agents' bodies in space and to the affordances of space in itself. Moreover, the physical space of the workshop allows authors to "enter" their story. By moving around the GREENHOUSE and arranging elements, they create a sense of both space and time that's difficult to achieve when writing on paper (Fig. 6.5). This spatialization helps authors visualize the "momentum" of the story and experience it from within as writer 3 sums up nicely his embodied experience:

- *I think here maybe we did it that way because we knew we had a locked space* [the emerging story takes place in a "closed" theatre space] *and this is a locked space* [in reference the 3D GREENHOUSE], *so it's perfect to recreate it, as if we could enter the story, I think… We spatialized the momentum of the story. Yeah, and it's not only… it's not a setting. We're in the story, precisely, time and space. And we can move in it, and you never do that, basically, when you're writing on the paper…you create time by walking around* (Fig. 6.5).

Discussion: Insight from the 4E Perspective

To shed light on LGO practices from the perspective of embodied cognition, we will start with first E, "Embodiment." In the context of these workshops, when writers are in ideation phase, either because they are in the very beginning of the digging process inside a project, or because they are in deep rethinking phase, the act of brainstorming with other group members while walking on limitless beeches potentially for hours has become key to unlocking possibilities. When the writer is involved (while walking) in telling what he/she sees while building a new story path, recording what comes out can prove very useful. It feels that the "telling while walking" unlocks a powerful driving force. As if the writer was walking inside his/her own story world, without being or feeling judged, as if in a sort of light trance or state of flow. This sense of being "carried out" designates a temporary loss of self-consciousness characteristic for the flow state (Csikszentmihalyi, 1996). Furthermore, unlike brainstorming seated around a table, it becomes clear that while walking and talking together, the brainstorming partners are helped by the fact that nobody is facing anybody. They all head toward somewhere (the imaginary place or world), walking side by side, as if the movement in space of each of them was helping the movement inside the story that is (re)discovered. Similarly, inside the GREENHOUSE, while heading toward a path between ideas, places, characters, actions, or trouble hanging from the ceiling, it is as if the movement of the body in space and the embodiment of a character were triggering this light trance or state of flow. Research has shown that as a writer "*gets into the skin of a character*," he or she often experiences intense moments in which intuition, unconscious, and spontaneous process take over the generation and selection of creative ideas (Bourgeois-Bougrine et al., 2014). The empathy with the characters leads to the feeling that they came to life and make the choices instead of the writer "*hijacking or guiding the story line in unexpected ways*" as reported by Australian fiction writers (Paton, 2012, p. 74).

Second E: "embedded cognition." Here, we can refer to tools and artifacts used to scaffold the writers' cognitive processes and to facilitate distributed cognition. These include (1) the rhizomatic keywords and drawings made in the early process of externalization, (2) the inspirational sources proposed to populate the GREENHOUSE, and (3) the affordances of the GREENHOUSE itself. The rhizomatic arborescence opens possibilities through word associations. Similarly, the drawings on paperboard of the world they "see" from what they hear create, as much consciously as subconsciously, possibilities for human situations or story paths. This is like what happens when acquiring knowledge through active manipulation using instruments instead of passive observation. Moreover, three main inspirational resources were used: a collection of face images specific, magazines for tearing out relevant visuals, and AI-generated images to spark unexpected ideas. Browsing peacefully among magazines, AI-generated images, and hundreds of pictures of human beings of various ages, races, shape spread on tables creates a clear feeling of encounter when the writers recognize an aspect of a human being that can feed their world, or another, there, half hidden under the scattered images, or a weird drawing facing them in a very unique manner. Using scissors, glue, clips, wool strings, etc., the writers collectively create collages, topographies, connections, networks of fears, characters, places, etc., which end up hanging somewhere in the GREENHOUSE. The decision of how to organize the space is intuitive, based on an impulsion, a movement of the body, a gesture that feels needed. The affordances of the GREENHOUSE environment itself plays an enabling role by allowing all these new elements to hang somewhere in the GREENHOUSE.

Third E: Extended cognition. We look at it as the actions employed by the writers in navigating and negotiating the artifacts they hanged in the GREENHOUSE to create surprises and generate possibilities. In our embodied experience, once the fictional world of the GREENHOUSE has been populated, in a spontaneous and intuitive way, using artifacts, people, places, shapes, obsessions, dreams, hanging like climbing plants, a second phase of creative exploration starts under dim light. Like speleologists, the writers navigate within the GREENHOUSE using a beam light to observe in detail, to scrutinize closely and from various perspectives (provoked by head or body movements) the hanging faces, drawings, collages, etc. Unexpected encounters or surprises can lead every writer to invent new paths, new links that can be materialized by a string of wool of different colors, depending on the nature of the link: love, hatred, jealousy, hope of a new start, manipulation. In term of situated cognition, the 3D physical space of GREENHOUSE facilitates "walking through their narrative" and connecting physically different elements in space and time.

Last but not least, Enacted cognition. It allows the writers to collectively generate a meaning through collaborative discussions, negotiations, actions, and behaviors. They make a sense of the story by allowing the hanging artifacts to be switched, changed from one side to another. Collectively, the writers experiment with rearranging story elements spatially to discover new narrative possibilities. They are thinking in action. The GREENHOUSE creates a common area where externalized ideas become shared property. This fosters natural collaboration as the writers built on and connect each other's contributions. The spatial arrangement of story elements help the writers visualize pacing, connections between scenes, and overall story arc in a more intuitive way. The GREENHOUSE allows a collaborative input and physical engagement with concepts before getting locked into a particular written form. Walter J. Ong (2012, pp. 77–81) famously proposed that "*writing restructures consciousness... Writing or script differs as such from speech in that it does not inevitably well up out of the unconscious.*" In his analysis of whether culture is oral or written, Declan Lloyd (2023, p. 316) delved deeper into Ong's theories and states that the oral mindset is "*strictly confined to the present... in orality word meanings come continuously out of the present... this is a mentality in a constant state of galvanized flux and dynamism, a state of pure impulsivity, a flurry of instant thoughts and acts and sensations....though words are grounded in oral speech, writing tyrannically locks them into a visual field forever.*"

Conclusion: Perspectives for Research and Future Developments

The GREENHOUSE consists of 3D physical narrative spaces with a mix of spatialized sensorial experiences, object interaction alongside a combination of augmented reality, artificial intelligence tools, and oral narration. The embodied experience in the GREENHOUSE enhances creative writing by encouraging authors to externalize ideas through visual and tactile methods, fostering collaboration and new perspectives, while addressing challenges of traditional writing processes. Initial tests, conducted at scale close to operational level, showed that the GREENHOUSE has the potential to accelerate both individual and collective creativity, thanks to the affordances of the setups, constant collaboration, and the pivotal role played by LGO's highly experienced coaches. Yet the "GREENHOUSE" outcome and performance need consolidation through new confirmatory testing.

Three hypotheses were advanced to explain how the "GREENHOUSE" supports ideas proliferation and cross-fertilization: (1) the 3D narrative space of the "GREENHOUSE" supports spatial and embodied cognition, (2) authentic dialogue and creative collaboration, between different stakeholders from various backgrounds, based on oral narration, and (3) populating the space with rich and diverse material helped "engineer" serendipity. Yet these hypotheses need verification. The 4E perspective allowed us to dig deeper in each E of the embodied experience despite some of the shortcomings of the rather opaque formulation of embodiment (Goldman & de Vignemont, 2009). Moreover, because of the convergence between the 4 Es, it was challenging to address them in separate ways without losing the holistic view of the process of story codevelopment. A more integrative model of the 4Es is needed in the case of the "creativity in the Wild."

To identify and describe the key mechanisms underlying the GREENHOUSE methodology we propose three research questions:

1. How spatial and embodied interaction facilitates deeper collective creative thinking?
2. How structured yet open-ended collaborative environments enhance creative dialogue and cocreation?
3. How diverse materials and multimodal stimuli help generate moments of serendipity and stimulate idea generation?

Answering these questions will involve the development of a valid conceptual framework and associated tools for real-world testing of the GREENHOUSE. This includes defining indicators of performance and designing evaluation protocols. Moreover, further elaboration of the creative output and the feedback of the testing phase are pivotal to make the GREENHOUSE concept replicable, as regards the integration of all its components such as the use of AI and VR or the automated collection and analysis of the spatial interaction and the creative collaboration. Indeed, the GREENHOUSE which was originally designed and tested in the context of script codevelopment for screenwriters and filmmakers has since attracted the interest of a larger community of creators like game designers, educators, policy-makers and foresight experts. Each of these groups seems to have specific needs and problems to solve. For example, in education, one of the major questions is "how to involve children and youth in today's societal issues (climate change, European democracy, etc.) through storytelling?" Organizations and businesses seek to develop a forward-looking or prospective approach for an economic or societal issue. This strategic foresight practice which relies on the development of different alternatives for future scenarios could be hampered by the "failure" of imagination. In response to this large interest, there is a need to systematically identify and categorize the needs, attributes, etc., of each group of user in order to expand the "GREENHOUSE" to these domains. Future developments will include the design of a virtual GREENHOUSE.

Acknowledgments The authors are grateful to the European program LIM—Less Is More, through its sub-program StoryTANK, supported by Creative Europe and would like to thank Aurélie Valat, Brigitte Bouillot, Vassili Silovic, and Lilo Malherbe for their support. The GREENHOUSE experience has been supported by the French Ministry of Culture through the France 2030 program.

References

Bachelard, G. (1994). *The poetics of space* (p. 150) (M. Jolas, Trans.). Beacon.

Bourgeois-Bougrine, S., & Glăveanu, V. P. (2018). Collaborative scriptwriting: Social and psychological factors. In *The creative process: Perspectives from multiple domains* (pp. 123–154).

Bourgeois-Bougrine, S., Glaveanu, V. P., Botella, M., Guillou, K., De Biasi, P. M., & Lubart, T. (2014). The creativity maze: Exploring creativity in screenplay writing. *Psychology of Aesthetics, Creativity, and the Arts, 8*(4), 384–399.

Burr, C. (2017). Embodied decisions and the predictive brain. In T. Metzinger & W. Wiese (Eds.), *Philosophy and predictive processing* (Vol. 7). MIND Group. https://doi.org/10.15502/9783958573086

Chatman, S. (1978). *Story and discourse: Narrative structure in fiction and film.* Cornell University Press.

Clark, A. (2018). A nice surprise? Predictive processing and the active pursuit of novelty. *Phenomenology and the Cognitive Sciences, 17*(3), 521–534.

Cole, M. (1996). *Cultural psychology: A once and future discipline.* Belknap Press.

Csikszentmihalyi, M. (1996). *Creativity: Flow and the psychology of discovery and invention.* Harper Collins.

Friston, K., Thornton, C., & Clark, A. (2012). Free-energy minimization and the dark-room problem. *Frontiers in Psychology, 3*(130), 1–7.

Fugate, J. M., Macrine, S. L., & Cipriano, C. (2019). The role of embodied cognition for transforming learning. *International Journal of School & Educational Psychology, 7*(4), 274–288.

Gallagher, S. (2005). *How the body shapes the mind.* Oxford University Press.

Getzels, J. W. (1979). Problem finding: A theoretical note. *Cognitive Science.*

Goldman, W. (1996). *Adventures in the screentrade: A personal view of Hollywood and screenwriting.* Abacus.

Goldman, A., & de Vignemont, F. (2009). Is social cognition embodied? *Trends in Cognitive Sciences, 13*(4), 154–159.

Green, M. C., & Brock, T. C. (2000). The role of transportation in the persuasiveness of public narratives. *Journal of Personality and Social Psychology, 79*, 401–421.

Kirchhoff, M. (2018). Predictive brains and embodied, enactive cognition: An introduction to the special issue. *Synthese, 195*(6), 2355–2366.

Krcmar, M., & Haberkorn, K. (2020). Mental representations. In *The international encyclopedia of media psychology* (pp. 1–17).

Kvernbekk, T., & Frimannsson, G. H. (2013). Narrative: A brief introduction. *Scandinavian Journal of Educational Research, 57*(6), 571–573.

Landrum, R. E., Brakke, K., & McCarthy, M. A. (2019). The pedagogical power of storytelling. *Scholarship of Teaching and Learning in Psychology, 5*(3), 247.

Lee, H., Bellana, B., & Chen, J. (2020). What can narratives tell us about the neural bases of human memory? *Current Opinion in Behavioral Sciences, 32*, 111–119.

Leung, A. K.-y., Kim, S., Polman, E., Ong, L. S., Qiu, L., Goncalo, J. A., & Sanchez-Burks, J. (2012). Embodied metaphors and creative 'acts'. *Psychological Science, 23*(5), 502–509.

Lloyd, D. (2023). The art of orality: How the absence of writing shapes the character of tribal, 'primitive' art. *World Art, 13*(3), 303–324. https://doi.org/10.1080/21500894.2023.2196250

Malinin, L. H. (2019). How radical is embodied creativity? Implications of 4E approaches for creativity research and teaching. *Frontiers in Psychology, 10*, 2372.

Marková, I. (2003). *Dialogicality and social representations: The dynamics of mind*. Cambridge University Press.

Ong, W. J. (with John Hartley). (2012). *Orality and literacy: 30th anniversary edition*. Taylor and Francis Group.

Oppezzo, M., & Schwartz, D. L. (2014). Give your ideas some legs: The positive effect of walking on creative thinking. *Journal of Experimental Psychology: Learning, Memory, and Cognition, 40*(4), 1142.

Paton, E. (2012). "When the book takes over": Creativity, the writing process and flow in Australian fiction writing. *The International Journal of Creativity & Problem Solving, 22*, 61–76.

Satama, S., Blomberg, A., & Warren, S. (2022). Exploring the embodied subtleties of collaborative creativity: What organisations can learn from dance. *Management Learning, 53*(2), 167–189.

Sawyer, R. K. (2006). *Explaining creativity: The science of human innovation by*. Oxford University Press. 368 pp. ISBN 0195304454.

Slaby, J. (2014). Emotions and the extended mind. In C. Scheve & M. Salmella (Eds.), *Collective emotions: Perspectives from psychology, philosophy, and sociology* (pp. 32–46). OUP.

Wang, S., Okada, T., & Takagi, K. (2023, October). How to effectively overcome fixation: A systematic review of fixation and defixation studies on the basis of fixation source and problem type. In *Frontiers in education* (Vol. 8, p. 1183025). Frontiers Media SA.

Wilson, A. D., & Golonka, S. (2013). Embodied cognition is not what you think it is. *Frontiers in Psychology, 4*, 58.

Zadbood, A., Chen, J., Leong, Y. C., Norman, K. A., & Hasson, U. (2017). How we transmit memories to other brains: Constructing shared neural representations via communication. *Cerebral Cortex, 27*(*10*), 4988–5000.

CHAPTER 7

Using a Large Volume of Local Historical Statistical Tables as a Source of Mutual Inspiration and Collective Creativity

Kumiyo Nakakoji, Yasuhiro Yamamoto, and Hibiki Sasaki

Introduction: Reimagining Historical Statistical Tables as Sources of Inspiration

Collective creativity emerges when individuals engage with shared artifacts, respond to each other's insights, and build layered understandings through mutual inspiration. This chapter explores how such creativity can be cultivated in a museum-like environment by reimagining a large volume of historical statistical tables from the Hakodate City History: Statistical Archives (1987) as cultural artifacts. These tables, long regarded as inaccessible and underused, are transformed into a medium for

K. Nakakoji (✉) • Y. Yamamoto
Future University Hakodate, Department of Media Architecture,
Hakodate, Japan
e-mail: kumiyo@acm.org; yxy@acm.org

H. Sasaki
Future University Hakodate, Graduate School of Systems Information Science,
Hakodate, Japan
e-mail: hsasaki.res@icloud.com

N. Bonnardel (ed.), *Collaborations in Creativity*, The Seven C's of Creativity, https://doi.org/10.1007/978-3-032-07753-0_7

citizen-driven inquiry and knowledge creation through the Stats.Hakodate project.

Local histories in Japan are often compiled into extensive multi-volume series, combining narrative texts with visual materials and statistical records. The Hakodate City History: Statistical Archives, which was published in 1987 as one volume in a ten-volume series, contains 712 statistical tables over 1260 pages, covering a broad spectrum of data from the 1860s to the 1970s (Fig. 7.1). The Statistical Archives is organized into three major thematic sections: *government*, *industry and economy*, and *society and education* (Table 7.1). This classification offers a useful lens for understanding the wide-ranging topics covered by the records, which span administrative systems, industrial output, demographic shifts, educational trends, and more.

As of 2020, while the narrative volumes had been digitized and made publicly accessible, this dense statistical record remained untouched. Two primary reasons seem to contribute to this lack of attention. The first is technical difficulty: the complexity of table layouts, which often extend

Fig. 7.1 Hakodate City History: Statistical Archives (published in 1987, covering the period from the 1860s to the 1970s) (Hakodate City, 1987)

Table 7.1 The contents of the Hakodate City History: Statistical Archives

Parts	*Topics covered*
Government	Meteorology, land, population, finance, elections
Industry & economy	Agriculture, fisheries, Russian Far Eastern fisheries, manufacturing, enterprises, commerce and finance, transport and communications, domestic trade, foreign trade, prices and wages
Society & education	Judiciary and police, customs, firefighting, sanitation, utilities (water, electricity, gas), education, military affairs

across multiple pages and include irregular merged cells and notes, renders machine recognition error-prone and inconsistent. Manual correction is unavoidable, making the process laborious and expensive. The second is a perceived lack of relevance: unlike narrative accounts that invite imaginative engagement with the past, statistical tables are often seen as dense and difficult to interpret, offering less immediate narrative appeal to general readers.

Through close examination, we found not only data but also anomalies—like a year in which every district reported the exact average household size. Other irregularities included a sharp rise in the number of recorded storm days in 1931, possibly due to a definitional change, and a concentration of police officers in one district, hinting at administrative restructuring (Nakakoji et al., 2020). These patterns triggered not confusion, but curiosity. They invited us to wonder.

These tables are more than passive data. They reflect layers of social structures, institutional practices, and even editorial intentions. They bear traces of past cultures and practices, such as now-defunct occupations, abolished tax categories, renamed districts, and unexplained omissions. These fragments offer glimpses into how the city once measured itself and what it chose to record or omit. Through this lens, statistical tables become more than information—they become cultural artifacts.

Hakodate, the city these tables represent, adds further richness. Located at the southern tip of Hokkaido with a current population of about 200,000, Hakodate was a major international trade port during the late nineteenth and early twentieth centuries. Its prosperity, driven by shipbuilding and fisheries, waned after repeated large-scale fires and economic restructuring. Today, it is primarily a tourist destination. The statistical records document this transformation in meticulous numerical detail, providing a unique opportunity to trace the rise and reconfiguration of a city.

These observations prompted us to reconsider the potential of historical statistical tables not merely as repositories of numerical data, but as cultural artifacts that can support engagement, interpretation, and collective reflection. This perspective led to the formulation of the Stats. Hakodate project, which explores how such materials might serve as a basis for participatory knowledge practices beyond conventional historical analysis.

This approach builds on prior research into inspirational museum environments (Yamamoto & Nakakoji, 2016), where mutual stimulation occurs not only through exhibits but through the circulation of externalized representations and asynchronous engagement among visitors. Drawing from design principles derived from museum fieldwork, such as encouraging experiential interaction, fostering peer-to-peer influence, and supporting cumulative expression, we developed two key components of this initiative: SLAPUP (Statistics Landscape Projections over Unseen Past), a series of participatory workshops in which participants generate annotations on printed or projected tables through discussion and reflection, and the Stats.Hakodate Browser, a digital platform that enables asynchronous engagement with these accumulated annotations.

Theoretical Foundation: Collective Creativity in Cultural Engagement

In this section, we introduce the theoretical foundation that informed our project. We draw on existing literature on creativity, with a particular focus on the concept of collective creativity and its relevance to artifact-mediated engagement. These ideas also inform how we interpret museums, not only as cultural spaces that house artifacts, but as participatory environments that support evolving relationships and shared meaning-making. This theoretical lens underpins our project's design approach, which is further discussed in the following section.

Collective Creativity

Creativity is not limited to moments of solitary invention or simultaneous collaboration. Instead, it arises through layered interactions with externalized representations—notes, diagrams, annotations, or even gestures—that persist in the environment and serve as stimuli for others (Nakakoji et al., 1998; Yamamoto & Nakakoji, 2005). Our approach focuses on *collective creativity*, which we define as "*the process and artifacts of individual creative knowledge work by interacting with 'knowledge in the world'—through external representations and communications with peers in a community*" (Nakakoji et al., 2000). Figure 7.2 illustrates this process, showing how individual expressions become part of a shared interpretive space to facilitate collective creativity.

Unlike social creativity, which focuses on the coproduction of collective outputs by groups or institutions (Fischer, 2004), collective creativity emphasizes the emergence of individual insights within a community context. It is rooted in asynchronous, distributed, and mediated interaction. In museums, a visitor's comment on a whiteboard or a sticky note attached to an exhibit can spark reflection by others, resulting in further elaboration

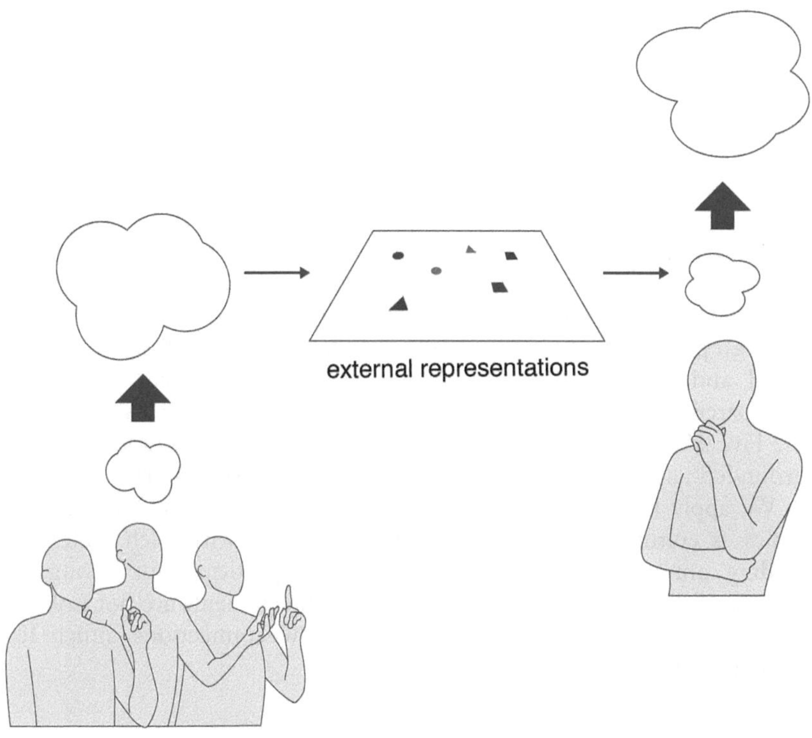

Fig. 7.2 Collective creativity mediated by external representations

or alternative interpretations. These interactions are often nonlinear and accumulate across time. This perspective is further supported by Maturana and Varela's (1987) notion of knowledge as an emergent property of interaction with externalized representations. The creative process, in this view, is a recursive engagement—between actors, representations, and evolving interpretations.

Our view of creativity aligns with the field of psychological creativity, rather than historical creativity (Boden, 1994). Our perspective aligns with the Four C Model of Creativity (Helfand et al., 2017), which distinguishes four levels of creative expression: Big-C (eminent achievements), Pro-C (professional-level innovation), Little-c (socially recognized everyday creativity), and Mini-c (personally meaningful insights not necessarily visible to others). The kinds of interpretive and reflective acts we observed—such as questioning anomalies in historical tables or proposing hypotheses—are best understood as expressions of Mini-c creativity, grounded in subjective meaning-making and situated engagement. In our context, creativity manifests in acts such as noticing inconsistencies in historical data, asking questions, or offering speculative interpretations—actions accessible to non-experts, grounded in lived experience and situated interaction.

This understanding resonates with Glăveanu's (2013) Five A's framework, which reconceptualizes creativity as emerging from the interplay of Actor, Action, Artifact, Audience, and Affordances. Our project exemplifies this model: citizens (Actors) engage in interpretive Actions mediated by statistical tables (Artifacts), interacting with one another (Audiences) through persistent annotations and workshop dialogue, all shaped by the social and material Affordances of the tools and spaces we design. Compared to more individual-centric models such as Rhodes's Four P's, the Five A's better captures the distributed, asynchronous, and materially grounded nature of the creative processes we seek to support.

We apply this framing to our reinterpretation of historical statistical tables as cultural artifacts. When citizens engage with such tables—by reading, annotating, or discussing them—they participate in an ongoing process of distributed sense-making. These engagements are not isolated; they accumulate, interact, and stimulate new connections, much like exhibits in a participatory museum.

How Museums Facilitate Collective Creativity

Museums have traditionally been viewed as institutions for the display and preservation of cultural artifacts. However, more recent perspectives emphasize their role as interactive spaces where diverse actors—visitors, curators, volunteers, artists, and researchers—engage in mutual inspiration and meaning-making (Nakakoji et al., 2016). Rather than sites of unidirectional knowledge transmission, museums are increasingly recognized as dynamic environments in which ideas, interpretations, and creative expressions circulate among individuals.

Museums are also communities—not in the narrow sense of shared membership, but as evolving constellations of participation, interpretation, and mutual learning. This view resonates with Lave and Wenger's (1991) concept of legitimate peripheral participation, which describes how individuals gradually move from peripheral to more central roles in a community by engaging with shared practices and knowledge.

In a museum context, this process is evident when a newcomer, observing others engaging with exhibits or leaving comments, begins to participate and contribute their own perspectives. Over time, as individuals gain confidence and familiarity, they deepen their engagement. Curators and volunteers likewise evolve through their interactions with audiences, sometimes reinterpreting their curatorial frameworks based on unexpected visitor responses (Nakakoji et al., 2016).

Such evolution occurs both in the individuals and in the community itself (Nakakoji et al., 2002). As people contribute new observations and interpretations, the knowledge landscape of the museum is enriched and transformed. This dynamic interplay among actors and representations enables the circulation of inspiration and supports long-term collective creativity.

We interpret our data engagement framework within this evolving community model. As individuals annotate statistical tables, others respond, expand, or reinterpret these annotations—layering new meanings onto historical data. The process is similar to how museum exhibits accumulate interpretations over time, resulting in a networked, living archive of inquiry. In this way, statistical tables become not only objects of analysis but also catalysts for community formation and shared reflection.

The next section builds on this perspective to examine how these principles informed the design of workshops and digital environments for engaging with the Hakodate City statistical archives.

Museum-Inspired Design for Inspiration and Collective Creativity

To bring the vision of statistical tables as interactive artifacts into reality, we revisited insights gained from our previous research on museum experiences. In earlier work, we conducted on-site investigations of museums in Japan and abroad to understand how inspirational environments are designed and how creativity emerges within them (Yamamoto & Nakakoji, 2016). The findings from these investigations have deeply informed our approach in the Stats.Hakodate project.

Museums are not only repositories of preserved knowledge but also dynamic spaces where individuals engage with artifacts, leave traces of thought, and encounter the perspectives of others. We learned that inspiration and creativity in museums often arise not from isolated exhibits, but from a web of interactions among people, artifacts, and the space itself. By reflecting on these patterns, we asked: What kinds of design elements support mutual inspiration and the emergence of collective creativity across time and participants? How can such mechanisms be translated into digital and workshop-based environments for engaging with historical statistical data?

In what follows, we first describe the patterns of inspiration and interaction identified in our prior museum studies. We then distill from these observations a set of design principles that guided the development of SLAPUP workshops (described in section "Designing User Experiences with Historical Statistical Tables") and the Stats.Hakodate Browser (described in section "SLAPUP: Statistics Landscape Projections over Unseen Past")—tools that allow historical statistical tables to be reimagined as media for intellectual exploration and collective meaning-making.

The Process of Inspiration and Collective Creativity

Knowledge and experiences in museums are not merely transmitted unidirectionally; instead, visitors actively engage with exhibits, leading to new discoveries and creative thinking. This process forms a circulating system of inspiration and collective creativity. Several interaction patterns have been observed:

(a) **Curators' exhibit designs inspire visitors.** Well-defined themes and visually engaging designs stimulate visitors' curiosity, making exhibits more accessible and compelling. Additionally, the integration of digital displays and interactive technologies enhances the experience, creating immersive learning environments that encourage deeper exploration.

(b) **Visitors' actions inspire other visitors.** When visitors see others engaging with an exhibit, their curiosity is often sparked, leading them to interact with the display as well. This phenomenon is particularly evident at The Exploratorium in San Francisco, where visitors' interactions with exhibits become an integral part of the exhibition itself, attracting and encouraging further participation.

(c) **Visitors' questions and comments inspire curators.** Questions posed by visitors introduce fresh perspectives that can lead curators to reconsider exhibition content or even influence future research directions. At the California Academy of Sciences, for instance, visitors' inquiries have prompted curators and researchers to explore new topics, demonstrating how audience engagement can drive intellectual discovery.

(d) **Exhibition designers inspire curators.** Collaboration between exhibition designers and curators fosters innovative ways of presenting scientific content, enriching the overall museum experience. At the Natural History Museum of Bern, an art director's perspective significantly influenced new visual strategies for exhibitions, showcasing how interdisciplinary cooperation can enhance exhibit design and storytelling.

These observations suggest that museums foster inspiration and collective creativity through three key mechanisms: (1) exhibit design, (2) interactions among stakeholders, and (3) the accumulation and circulation of visitors' externalized representations.

Design Principles for Facilitating Inspiration and Collective Creativity

To foster inspiration and collective creativity in museums, it is essential to design them not only as knowledge providers but also as spaces that continuously stimulate intellectual and creative engagement. The following three design principles support this goal:

(1) *Designing Exhibits as Experiential Spaces*

The first principle involves incorporating visually engaging and interactive elements into exhibits to create an environment where visitors can actively engage with the content. Instead of merely providing passive information, it is crucial to design mechanisms that encourage visitors to formulate their own questions and explore further.

(2) *Encouraging Interactions Among Stakeholders*

The second principle focuses on designing opportunities for curators, experts, and visitors to influence one another and facilitate the exchange of knowledge and ideas. Through exhibitions and workshops, it is necessary to create an environment where visitors can learn from each other.

(3) *Generating a Chain of Creation Through Visitors' Externalized Expressions and Records*

Rather than making the experience a one-time event, it is important to implement mechanisms that allow visitors to record and share their insights and creative expressions in both digital and physical forms. These externalized expressions should function as feedback loops that inspire subsequent creative activities by others, contributing to the development of collective creativity.

Circulating Inspiration and Collective Creativity as a Function of Museums

When these design principles are combined, museums function not only as spaces for providing knowledge but also as environments that generate a continuous chain of inspiration and collective creativity. In particular, when visitors take on an active role rather than remaining passive observers, the creative process is further strengthened.

In museums, the roles of inspirer and inspired are not fixed. Visitors, curators, artists, volunteers, and all other stakeholders both receive inspiration and inspire others in various ways (Nakakoji, 2016). For example, an expression created by one visitor in response to an exhibit might later inspire another visitor who encounters it afterward. This phenomenon is especially noticeable in participatory museums like the Exploratorium.

In this way, museums function as spaces where inspiration and collective creativity circulate across multiple layers, continuously fostering intellectual discoveries and creative thinking. Drawing on the functions and benefits of such museums and their associated design principles, our project has designed a user experience—specifically through the SLAPUP workshop formats and the Stats.Hakodate Browser—that enables individuals to engage creatively with historical statistical tables. The next section introduces the mechanisms and frameworks developed to facilitate data exploration and collective creativity in this context.

Designing User Experiences with Historical Statistical Tables

As discussed in the previous chapter, designing a museum as a community that fosters collective creativity requires three key elements: (1) designing exhibitions as experiential spaces, (2) promoting interactions among stakeholders, and (3) generating a chain of creativity through visitors' externalized expressions and records. This chapter explores new ways of engaging with historical statistical tables by applying these elements to enable an experiential approach to city history statistical tables as a museum.

Tables are a form of visual information representation developed by humanity, serving as a powerful method for organizing and understanding data. A table has a two-dimensional structure with horizontal rows and vertical columns, each assigned specific attributes. The intersections of rows and columns contain numerical or textual data corresponding to these attributes.

When viewing a table, people trace rows and columns, compare values, and discern trends. However, beyond merely extracting information, one can explore the historical and societal contexts behind the data, leading to knowledge creation. We refer to this approach—engaging with data by exploring its background and uncovering new insights—as *data engagement* (Nakakoji et al., 2021).

For our project, we used a partial digital dataset of Hakodate City History: Statistical Archives, which we manually digitized. First, each page of the statistical records was scanned into digital images. To improve table structure recognition accuracy, preprocessing steps such as adding grid lines and adjusting margins were performed. Then, a machine-learning program (Liu et al., 2022; Fujiwara & Terasawa, 2024) was used for table structure recognition, and commercial OCR software was employed for text data extraction. Since the output contained many recognition errors and numerous tables spanned multiple pages, manual correction and reconstruction were necessary to generate CSV files for each table.

This study aimed to promote data engagement with historical statistical tables by implementing two key elements: (1) the design and execution of a workshop-based user experience called *SLAPUP* and (2) the construction of an information environment, *Stats.Hakodate Browser*, to drive cyclic engagement.

SLAPUP (Statistics Landscape Projections over Unseen Past) is a series of workshops designed for citizens to interact with historical statistical tables and gain new insights. By engaging with tables as exhibits, participants first observe numbers and categories, which naturally lead to curiosity and the formation of questions. As they explore further, they begin to notice subtle inconsistencies or unexpected details within the data. These discoveries prompt them to infer possible background factors and underlying causes, encouraging deeper analytical thinking. Through this process, participants also engage in imagining the historical contexts and societal conditions reflected in the data, fostering a richer understanding of the past.

Stats.Hakodate Browser is a web-based data engagement environment that accumulates and shares externalized expressions (annotations and reflections) generated through SLAPUP workshops, associating them with the corresponding statistical tables as follows:

- Users record observations and questions derived from statistical tables.
- Annotations are linked to relevant sections of statistical tables for visualization.
- Insights are asynchronously shared among users, fostering collective experience.

Through this system, accumulated discoveries and discussions about statistical tables serve as catalysts for new insights. Thus, Stats.Hakodate Browser plays a role in forming a collaborative knowledge network mediated by statistical tables.

The following two sections provide details on SLAPUP and Stats. Hakodate Browser, respectively.

SLAPUP: Statistics Landscape Projections over Unseen Past

This project refers to citizen-participatory workshops in which individuals or groups observe statistical tables and verbalize or record their thoughts, discoveries, and questions, which we call SLAPUP (Statistics Landscape Projections over Unseen Past) meetings.

Various SLAPUP formats have been explored and tested to date. SLAPUP workshops encourage citizens to interpret statistical tables from their unique perspectives, envisioning the landscapes and unseen pasts revealed through the data. That is, by tracing the characters and numbers recorded in the tables, participants "read between the lines" to reconstruct historical contexts and backgrounds.

SLAPUP fosters dialogue among citizens, allowing them to compare and overlay their observations and insights on individual statistical tables. This interaction helps refine citizens' interpretations of statistical data, leading to a richer representation of the statistical table data space.

In this project, the process of engaging with statistical tables is instantiated, including the following cyclic steps (Fig. 7.3):

(1) Observing statistical tables alone or in groups, or experiencing numerical values through physical representations, to identify points of discrepancy or interest
(2) Contemplating or investigating these points
(3) Generating insights and discoveries
(4) Sharing and discussing observations with others in group settings
(5) Generating further insights

The goal of SLAPUP is to facilitate citizen engagement with statistical tables through this iterative sequence of activities. Designing effective SLAPUP workshops requires an environment that explicitly marks the points where participants felt discrepancies, records why they felt that way, and links these observations to the corresponding statistical tables, thereby fostering further discoveries. In this research, linking observations and discoveries to specific locations within tables is referred to as annotation.

Each individual SLAPUP workshop is termed a *SLAPUP Meeting* (*SUM*). We have prototyped and tested five SUM types to explore optimal session durations and workflows to date (Table 7.2). This section explains

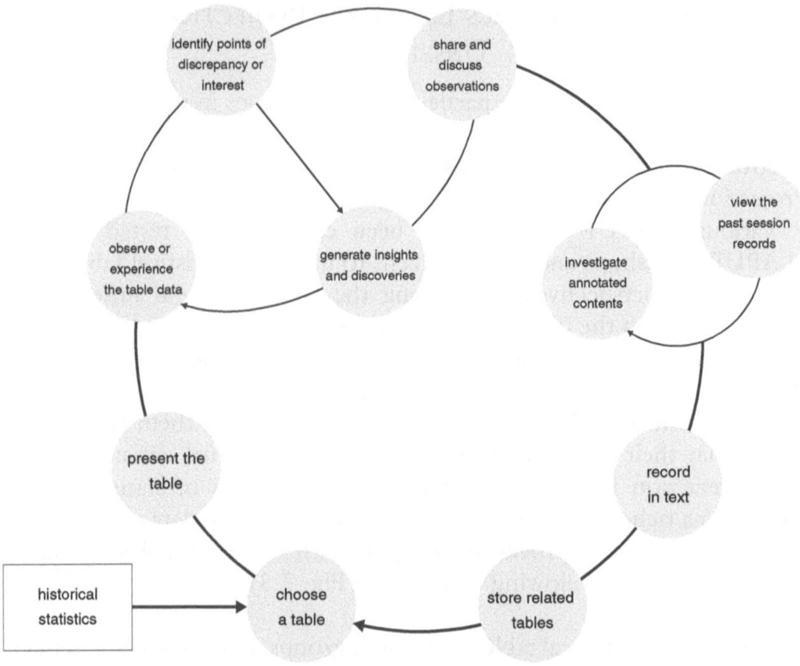

Fig. 7.3 A cyclic SLAPUP process. The diagram shows a series of concrete actions in the SLAPUP workflow, complementing the abstract five-stage cycle introduced in the text

each SUM type, discusses the findings from these trials, and considers the appropriate format for recording SUM sessions.

SUM Type 1: Solo Reading of Tables from a Physical Copy of the Book

Participants browse through the physical pages of the Statistical Archives and carefully examine the statistical tables that capture their interest. They mark noteworthy sections with small sticky notes and write down corresponding textual notes on a prepared whiteboard or notebook (Fig. 7.4).

This activity is conducted by a single participant, with an annotation session lasting approximately 15 minutes. In this format, participants actively select tables to explore. As a result, they can review multiple tables

Table 7.2 The five SLAPUP meeting (SUM) types

SUM Type 1	Solo reading of tables from a physical copy of the book 
SUM Type 2	Solo reading of a single table on an iPad 
SUM Type 3	Group reading of a single table projected on a large screen 
SUM Type 4	Group reading of a single table printed in a large format 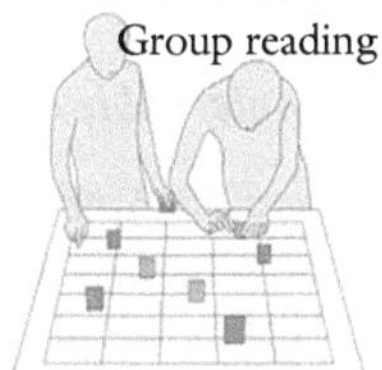
SUM Type 5	Solo reading of a single table by using DDrPP 

across different pages, tracing various contexts and making comparative observations during annotation. Additionally, engaging with the physical volume of the historical archives allows them to tangibly experience the accumulation of historical records through the texture and weight of the book.

Fig. 7.4 SUM type 1: solo reading of tables from a physical copy of the book (a digital clock on the table tracks elapsed time since the start of the workshop session. The same type of timer is visible in Figs. 7.5, 7.6, 7.7, and 7.9)

SUM Type 2: Solo Reading of a Single Table on an iPad

Each participant is provided with an iPad and an Apple Pencil. The application displays an image of a statistical table selected by the organizer, which the participant carefully examines. If they notice anything of interest, they directly annotate the table image using the Apple Pencil (Fig. 7.5).

There are two possible configurations: (1) all participants annotate the same statistical table or (2) each participant is assigned a different statistical table. In either case, the annotation session lasts approximately 15 minutes. Using pinch and swipe gestures, participants can adjust their view of the table and focus on specific points of interest. Additionally, the application's drawing tools allow for diverse annotation expressions, such as changing pen colors or using shapes to highlight certain data.

SUM Type 3: Group Reading of a Single Table Projected on a Large Screen

A statistical table image is projected onto a large whiteboard using a projector. A group of three to four participants examines the table together and writes down observations or points of interest using pens (Fig. 7.6).

Fig. 7.5 SUM type 2: solo reading of a single table on an iPad

The group engages in discussions and annotations while viewing the same statistical table. The annotation session typically lasts 20–25 minutes.

This format enables participants to grasp the overall structure of the statistical table and easily identify patterns in the data. Additionally, comparisons between distant rows and columns become more accessible, facilitating insights into relationships and correlations.

One drawback of projection-based display is that shadows may be cast depending on the participants' positions, making it difficult for some viewers to see the table clearly. Since participants write annotations in a large shared space, their work is fully visible to others. This visibility may serve as a constraint for some individuals who feel hesitant to write freely. Nonetheless, this format is well suited for collaborative engagement and shared learning.

SUM Type 4: Group Reading of a Single Table Printed in a Large Format

A statistical table image selected by the organizer is printed on an A0 (841 × 1189 mm) or A1 (594×841mm) panel and placed on a table. A group of three participants gathers around it to examine the table closely (Fig. 7.7). Participants write down their observations or questions on

Fig. 7.6 SUM type 3: group reading of a single table projected on a large screen

sticky notes, which they then attach to the corresponding locations on the panel. To minimize obstruction, the sticky notes are folded at the adhesive section so that they stand upright rather than covering the table itself.

Participants tag points of interest collectively, with an annotation session lasting approximately 15 minutes. Unlike SUM type 2 or SUM type 3, where annotations are directly written onto the table, this method relies on sticky notes as an intermediary medium. Unlike direct annotations—where numbers can be connected with lines or enclosed in rectangles—sticky notes make it more challenging to use visual elements such as diagrams or arrows. Instead, participants are encouraged to describe their observations in text as explicitly as possible.

Because sticky notes take up less space than direct annotations, they can serve as markers that point to relevant areas without cluttering the table. To enhance readability and ownership, sticky notes are color-coded by participant, making it easy to distinguish individual contributions. This method fosters an engaging and independent annotation process.

SUM Type 5: Solo Reading of a Single Table by Using DDrPP

DDrPP (Data Dressing and Prompted Physicality) is a set of data physicalization instruments developed for exploring the Statistical Archives of

Fig. 7.7 SUM type 4: group reading of a single table printed in a large format

Hakodate City History (Sasaki & Nakakoji, 2024). Data physicalization refers to the representation of numerical values and variations through geometric forms or material properties, leveraging human perceptual capabilities as a means of information transmission (Jansen et al., 2015).

DDrPP-Magnets is a tool that provides haptic feedback based on numerical values by utilizing the repulsive forces between magnets (Fig. 7.8). Users move a handle vertically, experiencing a push or pull sensation corresponding to the magnitude of the data. The device consists of a laser-cut MDF assembly housing magnets, servo motors, and an LCD screen. Neodymium magnets embedded in the top and bottom disks exert repulsive forces that adjust the pressure on the handle. A servo motor finely rotates the top disk, modulating the repulsive force and altering the perceived weight of the handle.

DDrPP-Magnets maps the spread of numerical values across years to variations in perceived weight. As users grasp the handle, they sequentially experience the physical sensation of data, with the LCD displaying the corresponding year. Prior user observation experiments suggest that broad trends in data can be perceived through haptic interaction (Sasaki & Nakakoji, 2024).

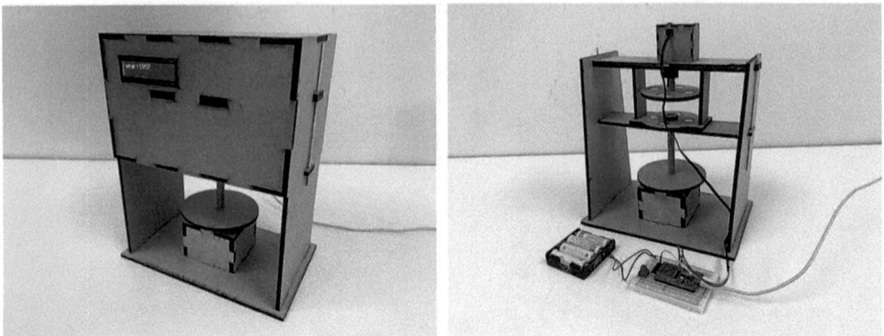

Fig. 7.8 DDrPP Magnets. Left: user-facing device with a round wooden disk, held to perceive statistical data as physical resistance. The display reads "year: 1957," indicating that the sensed value corresponds to a record from that year. Right: the internal mechanism, including magnets and electronic components

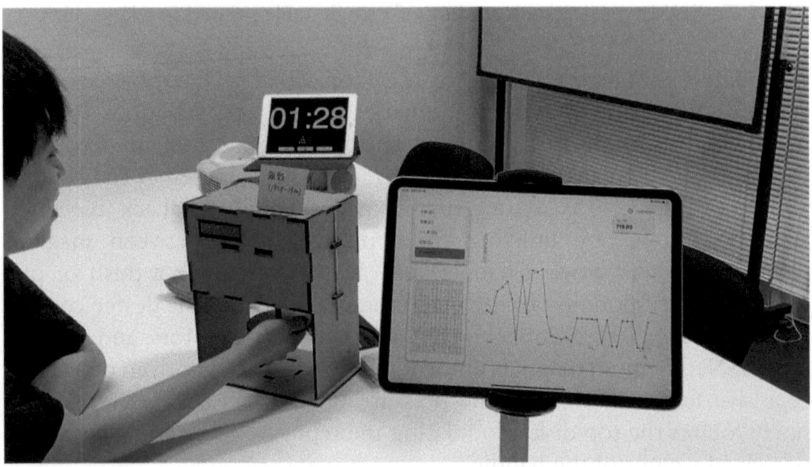

Fig. 7.9 SUM type 5: solo reading of a single table by using DDrPP

In SUM type 5, users experience the values of a statistical table selected by the organizer using DDrPP-Magnets (Fig. 7.9). For instance, they may perceive the fluctuations in grain production as physical weight changes, verbalizing their sensory impressions and data observations.

Two configurations were implemented: (1) a single participant verbalizing their impressions in real time and (2) an additional participant listening to and discussing the experience. In both cases, annotation sessions lasted approximately 15 minutes. Unlike visually browsing a statistical table, DDrPP-Magnets requires users to carefully process each numerical value through physical interaction. This approach allows users to detect subtle variations in seemingly uniform datasets. Furthermore, by associating haptic sensations with the historical background of the data, this method generates unique annotations distinct from visual exploration.

SUM Case Studies

A total of 13 SUM sessions (SUM001–SUM013) have been conducted to date. Table 7.3 summarizes each SUM session and its corresponding SUM type.

Each SUM followed a standardized protocol, where the organizer selected a statistical table in advance, and participants annotated notable points, following the cyclic process illustrated in Fig. 7.3. Each SUM session utilized one of the five SUM types described in the previous sections.

During trials, a stopwatch app on an iPad was used to measure session duration, while a separate iPad recorded video footage of the SUM process. Participants included students, educators, and external visiting researchers, all familiar with the history and geography of Hakodate City.

Recording Format of SUM Sessions

By systematically recording SUM sessions and linking annotations to statistical data, this project fosters cumulative knowledge creation through externalized expressions and documented experiences. Over time, this process enriches the statistical data space, transforming it into a multi-layered knowledge network (Fig. 7.10).

Currently, SUM annotations are manually transcribed into a standardized format (Fig. 7.11). Future improvements may integrate handwriting recognition, structured annotation linking, and speech recognition to enable semi-automated annotation documentation.

Stats.Hakodate Browser

Users participating in the SUM workshops view and annotate statistical tables to record their curiosities, questions, and inferred hypotheses. These annotations can then be asynchronously shared and collectively

Table 7.3 Case study—a summary of 13 SLAPUP meetings (the Page# column indicates on which page the statistical table used in the workshop is listed in the Statistical archives book)

SUM ID	*SUM type*	*Title of the statistical table used*	*Page# in the book*	*# of participants*	*Duration*	*# of annotations generated*
SUM001-1	SUM Type 2	*Hakodate Station Passenger and Freight Transport Status*	p.632	1	0:13:44	5
SUM001-2	SUM Type 2	*Hakodate Station Passenger and Freight Transport Status*	p.632	1	0:13:44	12
SUM001-3	SUM Type 2	*Hakodate Station Passenger and Freight Transport Status*	p.632	1	0:13:44	6
SUM002-1	SUM Type 2	*01: Imports by Item and Prefecture 1-1*	p.694	1	0:16:50	12
SUM002-2	SUM Type 2	*02: Production Volume and Value of Canned Salmon, Trout, and Crab*	p.495	1	0:16:50	7
SUM002-3	SUM Type 2	*03: Land Area by Land Use Category*	p.64	1	0:16:50	12
SUM003	SUM Type 3	*Household Census by Town during the Bunka and Bunsei Periods*	p.78	3	0:25:50	19
SUM004	SUM Type 5	*Cultivated Area of Major Crops (Postwar)—Rice*	p.404	1	0:03:32	17
SUM005	SUM Type 5	*Cultivated Area of Major Crops (Postwar)—Other Grains*	p.404	1	0:02:30	22

SUM006	SUM Type 5	*Cultivated Area of Major Crops (Postwar)—Legumes*	p.404	1	0:02:29	17
SUM007	SUM Type 5	*Cultivated Area of Major Crops (Postwar)—Oats*	p.404	1+1	0:02:18	45
SUM008	SUM Type 5	*Cultivated Area of Major Crops (Postwar)—Wheat and Barley*	p.404	1+1	0:02:06	51
SUM009	SUM Type 4	*Number of Fire Incidents 1* *Number of Fire Incidents 2*	p.1102 p.1103	2	0:17:25	25
SUM010	SUM Type 1	*Elections*	pp.366–395	1	0:15:21	6
SUM011-1	SUM Type 4	*Household Census by Town during the Meiji Period*	pp.86–87	3	0:15:20	37
SUM011-2	SUM Type 4	*Occupational Census of Hakodate City, Meiji 11 (1878)*	pp.92–93	3	0:15:20	32
SUM012	SUM Type 3	*Hakodate City Tram Operation Status*	p.637	3	0:20:40	23
SUM013-1	SUM Type 4	*University (New System)*	p.1196	3	0:14:17	16
SUM013-2	SUM Type 4	*Comprehensive Directory of Senior High Schools (New System)*	p.1195	3	0:17:42	14
SUM013-3	SUM Type 4	*Comprehensive Directory of Junior High Schools (New System)*	p.1194	3	0:17:56	12

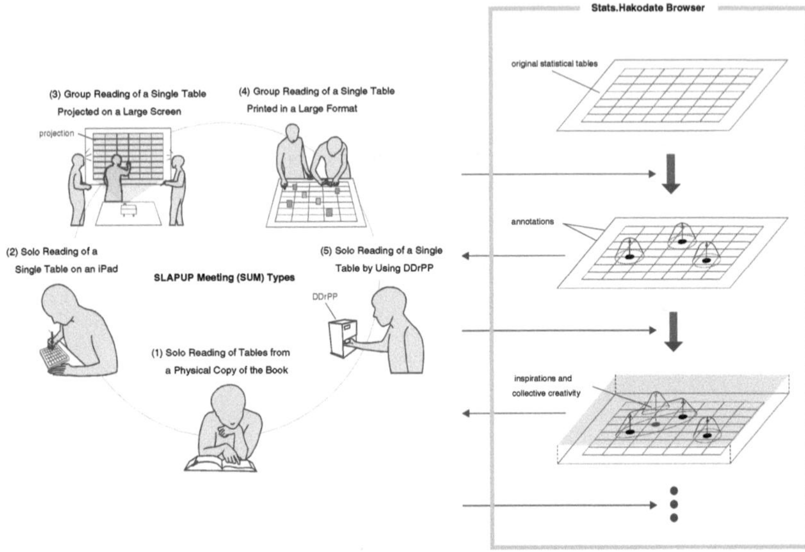

Fig. 7.10 Progressive enrichment and deepening of the statistical archive data space

experienced through the statistical tables presented in *Stats.Hakodate Browser* (Fig. 7.12).

Stats.Hakodate Browser enables users to view statistical tables one at a time while replaying annotations made by past users to the displayed table. When the annotation display mode is activated, a control bar appears below the table. By gradually dragging the handle to the right, annotations are replayed in chronological order according to when they were added (Fig. 7.13).

A screenshot from the Stats.Hakodate Browser shows a statistical table with overlaid handwritten annotations as part of the annotation playback interface (Fig. 7.13). Below the table is a horizontal playbar, which allows users to replay annotations in chronological order by dragging a slider from left to right.

Annotations are indicated by small red circular markers displayed on the statistical tables. The positions of these red circles correspond to locations where SLAPUP workshop participants added annotations. When users hover the mouse over a red circle, a pop-up displays the content of the annotation.

sum001A

SUM Type: 2

Summary:
Three participants annotated the same statistical table on their individual iPads using the photo editing function. Afterward, they shared their findings with each other.

Date: [redacted]

Time: 15:38

Duration: 20 minutes 25 seconds (Annotation: 14 minutes, Sharing: 6 minutes)

Place: FUN R724

StatsBookPage: p.632

Stats Table Title: Hakodate Station Passenger and Freight Transport Status

Annotator: [redacted]

Device Used: iPad (8th generation)

Device Used: Apple Pencil

Stats File Type: PDF

Annotations

AT01: The number of passengers boarding is always higher.
AT02: [Blue highlight] A significant decrease compared to the previous year?
AT03: [Red highlight] A significant increase compared to the previous year?
AT04: Does this indicate the total of both inbound and outbound traffic?
AT05: The number of arriving "inbound" passengers is large.
AT06: The number of departing "outbound" shipments is large.

Notes

SUM001A_AT02-03: The increase in passengers does not correlate with population growth. While the population fluctuated, the number of passengers continued to increase. This suggests that there may have been reasons compelling people to leave Hakodate, such as labor migration or evacuation. (*Reference: Hakodate City History: Statistical Archives, pp. 72-74, Household and Population 1-1, 1-2*)

SUM001A_AT04: On July 31, 1924, freight transport was first initiated along the entire Japan Sea Trunk Line. However, since the Seikan Tunnel had not yet opened at that time, it is unlikely that railway freight transport had a significant impact.
(*Reference: Japan Freight Railway Company, 150 Years of Freight Railway Transport History, URL: https:// www.jrfreight.co.jp/event150/history.html*)

However, on August 1, 1925 (Taisho 14), vehicle ferry transport was introduced, and until the opening of the Seikan Tunnel, the Seikan Ferry was the dominant transport method. Additionally, seafood from Hokkaido began to be transported as far as the Kanto and Hokuriku regions, leading to what was called a distribution revolution.
(*Reference: Hakodate City Seikan Ferry Memorial Ship Mashu-Maru, History of the Seikan Ferry, URL: https://mashumaru.com/?%E9%9D%92%E5%87%BD%E9%80%A3%E7%B5%A1%E8%88%B9%E3%81%AE%E6%AD%B4%E5%8F%B2,24*)

Fig. 7.11 An example of recorded annotations (from SUM001A; the original is written in Japanese)

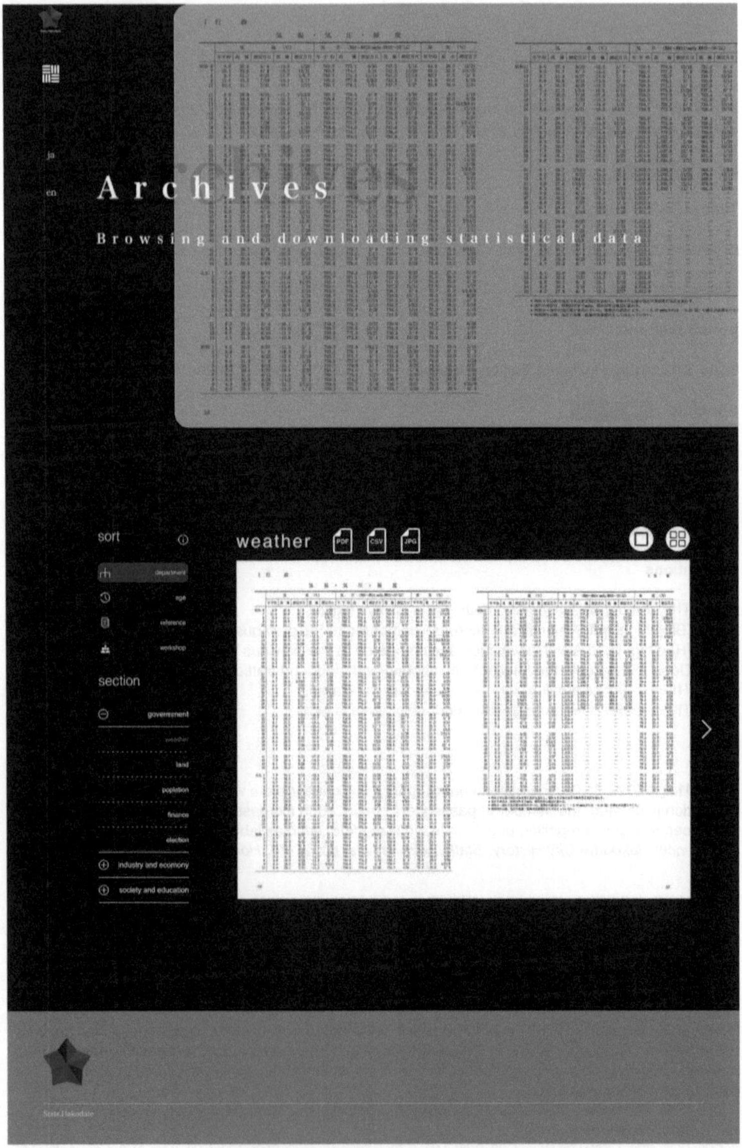

Fig. 7.12 A screenshot of Stats.Hakodate Browser

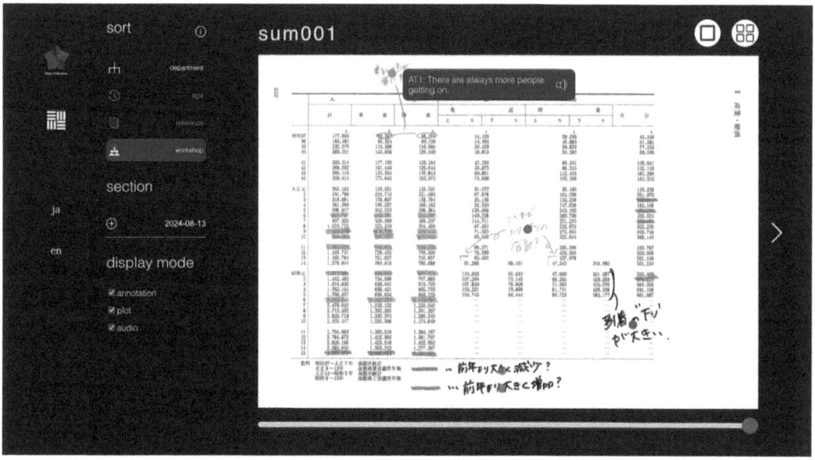

Fig. 7.13 Replaying the annotation process for the table

Stats.Hakodate Browser implements two modes for annotation displays.

(1) **Annotation Discovery Mode.** This mode displays red circular markers at the locations where annotations exist, indicating only the presence of annotations. It is designed to help users discover where annotations have been made within the statistical tables. The red circles suggest that "something of interest is here," encouraging users to engage in an exploratory experience.

In this mode, users can only see that a particular area of the table has been annotated without revealing the content of the annotations. This allows users to ask themselves, "Is there something interesting in this section?" and explore freely. By hovering the mouse over a red circle, they can view the corresponding annotation content, effectively "checking their answers" against the perspectives of others.

(2) **Annotation Playback Mode.** This mode overlays recorded annotations directly onto the statistical table. It functions as a playback mode that replays how past annotations were added in chronological order, allowing users to trace the thought processes of previous annotators and deepen their understanding of the statistical tables.

Since annotations are displayed directly on the tables, users can reconstruct the contextual background in which the annotations were originally made. If an annotation applies not just to a single cell but to an entire column or spans multiple cells, it may be difficult to interpret the annotation's intention based solely on its textual content. By replaying the annotation process in chronological order, users can relive the context in which annotations were made, making the experience more vivid and immersive.

This functionality is similar to how museum visitors become intrigued by observing others interacting with exhibits. It allows users to retrace the moments of surprise and discovery experienced by past SLAPUP workshop participants. As a result, users do not simply browse the data but engage in a dialogue with the perspectives and thought processes of others, enhancing their understanding of the statistical tables.

Musealization of Historical Statistical Tables as a Foundation for Collective Creativity

This study has reinterpreted the Hakodate City History: Statistical Archives not as a passive accumulation of historical data, but as a public medium for inspiration, dialogue, and the generation of new questions and insights. Through the design of SLAPUP workshops and the Stats. Hakodate Browser, we have transformed historical statistical tables into interactive artifacts—objects that invite citizens to engage in acts of noticing, wondering, annotating, and sharing. These tables no longer function merely as static records of the past but have become a dynamic platform for collective creativity.

One of the key insights from this project is that fixed data can support evolving, plural interpretations. While the numerical values remain unchanged, their significance shifts and multiplies as citizens engage from different perspectives. This engagement leaves visible traces in the form of annotations—persistent representations that accumulate, intertwine, and invite response. These traces do not simply document reactions; they generate chains of reflection and reinterpretation.

In total, over a dozen workshop sessions produced more than 250 annotations. Among these, 238 were analyzed and categorized into 82 distinct interpretive codes, grouped into 12 thematic clusters using qualitative methods such as the KJ technique (Sasaki & Nakakoji, 2025). These annotations ranged from simple observations to historically informed hypotheses. For instance, one participant suggested that a spike in police officers in a certain district might indicate the location of government housing. Another inferred that a change in how passengers were counted at Hakodate Station marked its shift from terminus to transit station. These examples highlight how data became a stimulus for collective sense-making.

Such annotations are not isolated. They spark conversation, provoke new questions, and prompt others to revisit the data from fresh angles. We observed these dynamics during the in-person workshops, where participants built on each other's remarks, moved across the table's structure, and recursively refined their interpretations through dialogue. The annotations generated in this context served as shared externalizations that facilitated collective sense-making within the workshop setting.

While these collaborative dynamics have thus far emerged primarily through synchronous, face-to-face interaction, the Stats.Hakodate Browser is designed to extend this potential into asynchronous environments. Although the current data does not yet reflect long-term, distributed annotation-based dialogue, we see the browser as an infrastructure that could support time-shifted mutual inspiration and creative reinterpretation in the future.

We interpret this as a grounded, cumulative form of collective creativity: not co-creation in a conventional sense, but an ongoing ecology of shared noticing, response, and reinterpretation. Participants do not necessarily create something together in real time, but their insights, once externalized, become part of a shared intellectual landscape that others can enter, navigate, and expand.

Importantly, this approach goes beyond making data more "accessible" or "searchable." It reimagines the statistical archive as a participatory cultural infrastructure—a space for reflecting on the past and imagining the present through multiple lenses. This vision aligns with recent shifts in public humanities (Smulyan, 2022) and digital heritage that emphasize co-interpretation, dialogic engagement, and locally rooted knowledge practices.

Looking forward, several avenues warrant further exploration. First, expanding the expressive range of annotation modalities (e.g., audio, sketches, gestures) could invite broader forms of participation. Second, refining the Stats.Hakodate Browser's interface and interaction design will be essential to supporting more accessible engagement. Third, applying this musealization approach to other local historical datasets may offer opportunities for comparative analysis and foster a distributed civic network for reinterpreting regional archives through creative, reflective interaction.

This work illustrates a form of creativity that emerges through a combination of human-human dialogue and digital mediation. While collaborative interpretation unfolded primarily through in-person workshops, our tools also suggest future directions for supporting creative engagement across time and contexts. In this way, the project explores how cultural data environments might foster new modes of shared inquiry that are not limited to colocated or simultaneous collaboration.

References

Boden, M. A. (1994). *Dimensions of creativity.* MIT Press.

Fischer, G. (2004). Social creativity: Turning barriers into opportunities for collaborative design. In *Proceedings of the eighth conference on Participatory design: Artful integration: Interweaving media, materials and practices – Volume 1 (PDC '04)* (pp. 152–161). Association for Computing Machinery. https://doi.org/10.1145/1011870.1011898

Fujiwara, S., & Terasawa, K. (2024). *Table structure recognition with source-target attention.* IW-FCV.

Glăveanu, V. P. (2013). Rewriting the language of creativity: The five A's framework. *Review of General Psychology, 17*(1), 69–81. https://doi.org/10.1037/a0029528

Hakodate City History Editorial Office (Ed.). (1987). *Hakodate City history: Statistical archives.* (in Japanese).

Helfand, M., Kaufman, J. C., & Beghetto, R. A. (2017). The four C model of creativity: Culture and context. In V. P. Glăveanu (Ed.), *Palgrave handbook of creativity and culture research* (pp. 15–360). Palgrave.

Jansen, Y., Dragicevic, P., Isenberg, P., Alexander, J., Karnik, A., Kildal, J., Subramanian, S., & Hornbæk, K. (2015). Opportunities and challenges for data physicalization. In *Proceedings of the 33rd annual ACM conference on human factors in computing systems (CHI '15)* (pp. 3227–3236). Association for Computing Machinery. https://doi.org/10.1145/2702123.2702180

Lave, J., & Wenger, E. (1991). *Situated learning: Legitimate peripheral participation.* Cambridge University Press.

Liu, H., Li, X., Liu, B., Jiang, D., Liu, Y., & Ren, B. (2022). Neural collaborative graph machines for table structure recognition. In *Proceedings of the IEEE/CVF conference on computer vision and pattern recognition (CVPR 2022)*.

Maturana, H. R., & Varela, F. J. (1987). *The tree of knowledge: The biological roots of human understanding*. Shambhala Publications.

Nakakoji, K. (2016). Chap. 2: Issues and challenges for instrumenting inspirational experiences at museums. In *Inspirational museums: Toward new horizons for cultural public spaces* (pp. 11–30). Airi Shuppan Co. (in Japanese).

Nakakoji, K., Ohira, M., & Yamamoto, Y. (2000). Computational support for collective creativity. *Knowledge-Based Systems, 13*(7–8), 451–458. https://doi.org/10.1016/S0950-7051(00)00066-1

Nakakoji, K., Shindo, H., Yamamoto, Y., & Okada, T. (Eds.). (2016). *Inspirational museums: Toward new horizons for cultural public spaces*. Airi Shuppan Co. (in Japanese).

Nakakoji, K., Yamamoto, Y., & Matsubara, N. (2021). Designing data interactivity for visualizing data. *The Journal of the Institute of Electronics, Information and Communication Engineers, IEICE*, 197–205. (in Japanese).

Nakakoji, K., Yamamoto, Y., Matsubara, N., Kawashima, T., & Kimura, K. (2020). Toward digitalizing historical statistical tables of Hakodate. *Journal of Japan Society of Information and Knowledge, 30*(2), 176–181. https://doi.org/10.11501/11750989. (in Japanese).

Nakakoji, K., Yamamoto, Y., Nishinaka, Y., Kishida, K., & Ye, Y. (2002). Evolution patterns of open-source software systems and communities. In *Proceedings of the international workshop on principles of software evolution (IWPSE '02)* (pp. 76–85). Association for Computing Machinery. https://doi.org/10.1145/512035.512055

Nakakoji, K., Yamamoto, Y., Suzuki, T., Takada, S., & Gross, M. D. (1998). Beyond critiquing: Using representational talkback to elicit design intention. *Knowledge-Based Systems, 11*(7–8), 457–468. https://doi.org/10.1016/S0950-7051(98)00041-4

Sasaki, H., & Nakakoji, K. (2024). Encouraging citizens to actively engage in historical local statistical data through innovative interactive data representations. *Journal of Japan Society of Information and Knowledge, 34*(2), 190–195. https://doi.org/10.2964/jsik_2024_015. (in Japanese).

Sasaki, H., & Nakakoji, K. (2025). Analyzing and reflecting on participant annotations in workshops for reading local historical statistical tables. *Journal of Japan Society of Information and Knowledge, 35*(2), 1–6. (in Japanese) (in print).

Smulyan, S. (2022). Why public humanities. *Daedalus, 151*(3), 124–137. https://doi.org/10.1162/daed_a_01955

Yamamoto, Y., & Nakakoji, K. (2005). Interaction design of tools for fostering creativity in the early stages of information design. *International Journal of Human-Computer Studies, 63*(4–5), 513–535. https://doi.org/10.1016/j.ijhcs.2005.04.020

Yamamoto, Y., & Nakakoji, K. (2016). Chap. 3: Studying inspirational exhibition designs and experiences at museums. In *Inspirational museums: Toward new horizons for cultural public spaces* (pp. 32–48). Airi Shuppan Co. (in Japanese).

CHAPTER 8

Human-AI and Social Robots Creative Collaboration

Sudapa Chompunuch, Florent Vinchon, and Todd Lubart

Introduction

Over the past few decades, artificial intelligence (AI) systems have moved from simple creative support tools (Shneiderman, 2002, 2007) to co-creators (Davis et al., 2015; Karimi et al., 2018; Rezwana & Maher, 2023) and even to become the principal creators (Colton, 2012; Colton et al., 2015; Elgammal, 2017; Mazzone & Elgammal, 2019) in the creative process. In the early creative stages, AI served human creativity, helping to perform functions like grammar checking, pattern recognition, or style recommendation, thus effectively serving as a digital assistant.

The development of machine learning and deep neural networks and the coming age of generative artificial intelligence (GenAI) has allowed AI

S. Chompunuch (✉)
Bangkok University, The Institute for Knowledge and Innovation Southeast Asia (IKI-SEA), Bangkok, Thailand
e-mail: sudapa.chom@gmail.com

F. Vinchon • T. Lubart
Université Paris Cité and Univ Gustave Eiffel, LaPEA, Boulogne-Billancourt, France
e-mail: florent.vinchon@gmail.com; todd.lubart@u-paris.fr

N. Bonnardel (ed.), *Collaborations in Creativity*, The Seven C's of Creativity, https://doi.org/10.1007/978-3-032-07753-0_8

to go beyond these simple tasks and processes, allowing it to generate entirely new content and new ideas. In this development of AI capabilities, rule-based interactions have been replaced by dynamic approaches in which AI can learn from large pre-trained models or large datasets and generate outputs that are new and relevant to human collaborators. As a result, AI has even transformed from mere tools or facilitators of human creativity to partners in the creative process. This also opens up future opportunities for collaboration in various fields, from arts to sciences.

Building upon this transformation, many scholars have identified several ways humans and AI can collaborate on creative processes. Most studies often presented three core modes or levels of human-AI collaboration (e.g., Davis et al., 2015; Karimi et al., 2018; Lubart et al., 2021; Zhang et al., 2023; Mao et al., 2024). While their terminologies and specific emphases are varied, the core idea is that humans and AI contributions can range from AI as a Supportive tool (often seen as a Helper), AI as a Principal creator/Creative agent, to AI as a Partner or Co-creator, depending on the level of AI autonomy and Human involvement. These three modes emphasize how shifts in autonomy and collaborative depth can define the creative relationship between humans and AIs.

The increasing sophistication of AI tools has expanded its role from augmenting small creative tasks to orchestrating complete creative workflows. Today, GenAI already has the ability to generate artwork, compose music, or write narrative stories that enhance the human creative process. Consequently, there is a growing interest among artists and innovators who are now using AI not only as a tool but also as co-creators that expand the limits of traditional artistic practices (Wingström et al., 2024). By studying this progression, we gain insight into how AI can be integrated in ways that enhance rather than diminish ***human-centered creativity*** and how human creators may increasingly leverage AI as a powerful partner that helps them push artistic and innovative boundaries.

Existing Frameworks and Levels of Human-AI Collaboration

In this section, existing frameworks and levels of human-AI collaboration are examined (from Davis et al., 2015; Karimi et al., 2018; Lubart et al., 2021; Zhang et al., 2023; Mao et al., 2024). AI contributions can range from AI as a supportive tool (often seen as a Helper) and AI as a principal creator/creative agent to AI as a partner or co-creator, depending on the level of AI autonomy and Human involvement.

For example, Davis (2015) and Karimi (2018) highlighted three main trends in creative systems. Creative systems are "intelligent systems that can perform creative tasks alone or in collaboration." The creative systems can be categorized into three main categories: Creativity support tools, Co-creativity, and Fully autonomous systems.

- **Creativity support tools** are tools and apps that are built to support user creativity (Shneiderman, 2007); in our chapter, creativity support tools can be referred to as AI that can support human creativity.
- **Co-creativity exists** when computers and humans collaborate with each other to build shared creative artifacts (e.g., Davis et al., 2015; Rezwana & Maher, 2023; Vinchon et al., 2023). Karimi (2018) describes a co-creativity system as the "Interaction between at least one AI agent and at least one human where they take action based on the response of their partner and their own conceptualization of creativity during the co-creative task" (Karimi et al., 2018). They also highlight a similar term called "Mixed initiative co-creativity" (Yannakakis et al., 2014). In their definition, "both the human and the computer take initiative in creating a new artifact, meaning both parties are actively contributing to the shared artifact." So, in a mixed-initiative system, the computational agent(s) contributes proactively rather than solely in response to a user request. The human and computational/AI agents do not need to contribute to the same degree, and there is no need for their contribution to be symmetrical.
- **Fully autonomous systems**. This category refers to AI as a principal creator or Creative agent. In our case, AI systems are built to generate creative artifacts that later are judged by humans to be creative (Colton, 2012; Colton et al., 2015; Elgammal, 2017; Mazzone & Elgammal, 2019).

Two other contributions from Zhang et al. (2023) and Mao et al. (2024) highlight the roles of AI in the creative process. Zhang et al. (2023) proposed a three-level framework that explains how AI should be involved in collaboration during the creative process.

- **Level 1, with low AI involvement**: Humans performing most of the stages, and AI should provide assistance or act as a supportive tool for each part of the creative process
- **Level 2, with moderate AI involvement**: Humans and AI engage in communication to exchange information (Zhang et al., 2021) and provide mutual guidance (Oh et al., 2018), and AI helps build and test the creative outcome.
- **Level 3, with the highest AI involvement**: "AI should be a more autonomous process, with AI participating in most of the stages based on requirements or inputs provided by humans" (Zhang et al., 2023).

Mao et al. (2024) examined the hybrid intelligence technology acceptance model (HI-TAM), and they identified four options for human versus AI involvement in the creative process. The AI roles include **AI generates for You, AI Co-Generates with You, AI Advises You, and No AI Involvement** (Mao et al., 2024).

Additionally, in Human-Robot Interaction (HRI) research, Lubart et al. (2021) focus on human creativity or human-centered HRI and propose three modes of human-robot co-creation, including **(1) robots as a support for human creativity, (2) humans as a support for robotic creativity, and (3) human-robot (humbot) co-creation.**

Drawing from the existing frameworks and levels of Human-AI collaboration, we proposed an overview of the three modes of Human-AI creative collaboration. These modes illustrate how responsibilities, control, and creative agency can shift between humans and AI systems.

Figure 8.1 illustrates three different modes of human and AI collaboration by using two color bars (**black** for how much human is involved and **gray** for how much AI is involved). In Mode 1, the black bar is larger, indicating a "human-centered" approach where AI serves mainly as a helper. Mode 2 shows the black and gray bars roughly balanced, indicating equal participation or "co-creation" between humans and AI. In mode 3, the gray bar is dominant, indicating a scenario in which the AI takes on the principal role while human involvement is minimal. In case the highest involvement of pure humans is 100%, it means no collaboration; if there is 100% AI involvement and 0% human involvement, that means there is also no collaboration between the two parties. This latter case is purely theoretical. Sometimes, even if AI did "all the work," humans will still be involved in the "reception phase" by interpreting the AI's work and

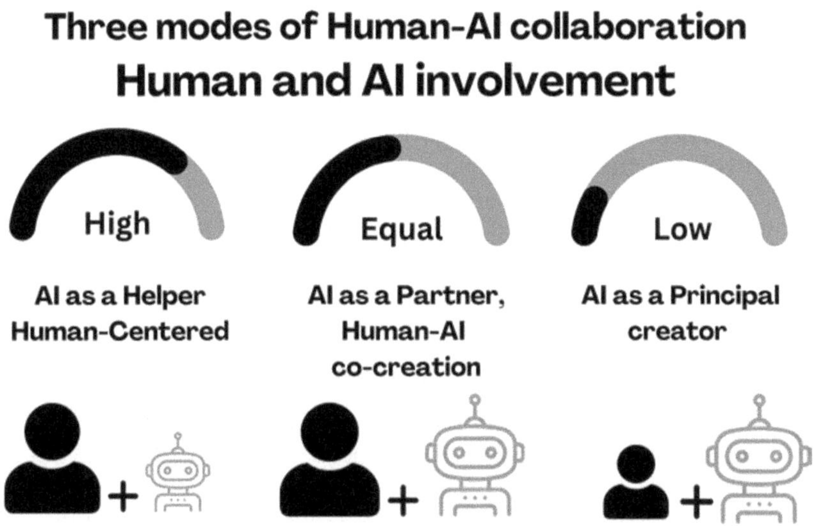

Fig. 8.1 The involvement of humans and AI in three modes of human-AI collaboration

recognizing its value. In general, from the social constructionist point of view, in Actor-Network Theory (ANT), the artifact or the tool can be viewed as an "actor," suggesting that tools are naturally collaborating with humans in the way (Latour, 2007; Glaveanu et al., 2020).

Mode 1: Human-Centered Creativity Aided by AI (AI as a Helper)

- **Definition**: AI primarily acts as a support tool or facilitator for each creative stage.
- **Human Role**: Humans remain the main creators and drive most of the creative process. They originate ideas, make final decisions, and shape the creative vision.
- **AI Role**: AI as Nanny, Pen-pal, Coach (Lubart, 2005). The AI supports human creators in the creative process, such as problem identification, idea generation, idea selection and evaluation, and the implementation of ideas.

Mode 2: Human-AI Co-creation (AI as a Partner)

- **Definition**: Humans and AIs engage in iterative co-creation processes, where both parties influence each other's contributions over time, leading to more synergistic creation processes (humans and AI work together back and forth, make improvements together, and help refine content in turn).
- **Human Role**: Humans and AI continuously exchange ideas, iterating and building on each other's contributions in real time.
- **AI Role**: AI as Colleague, Partner, or Co-creator (Lubart, 2005).

Mode 3: AI as Principal Creator (AI-Centered)

- **Definition**: The AI takes a more proactive or generative role in creating ideas or outputs, while humans step in to guide, critique, and refine.
- **Human Role**: The human becomes a supervisor or mentor, offering feedback and setting constraints and guidance to ensure the AI's creative outputs align with desired goals or values.
- **AI Role**: AI as Creator or Creative Agent. AI leads the creative process or production using generative techniques (e.g., machine-generated paintings, musical compositions, or text) and refining outputs (post-editing) based on human input.

Three Modes of Human-AI Creative Collaboration

In this section, we discuss each mode of human-AI collaboration, then illustrate them with examples.

Mode 1

Human-Centered Creativity Aided by AI (AI as a Helper)

Views on Human-Centered creativity can be found in the works of Dautenhahn (2007) and Lubart et al. (2021), who examined Human-Robot interaction. The approach is centered on human-robot interaction (HRI), where AI and social robots were designed to interact with humans engaged in creative tasks, providing some input and stimuli to aid human creativity (Lubart et al., 2021).

In this chapter, we refer to "a social robot" as a machine that is equipped with artificial intelligence capabilities, allowing it to perform tasks that require some level of decision-making, learning, and adaptation to its environment. It essentially mimics human intelligence in its actions, rather than just following pre-programmed instructions; thus, the AI acts as the "brain" of the robot, enabling it to go beyond simple automation and react to complex situations (Doncieux et al., 2022). For example, in storytelling, a robot might provide cues or questions to stimulate further creative thinking. This interaction can help individuals explore new directions in their creative work (Lubart et al., 2021). Throughout this creative process, humans maintain full control over the creative vision and final decisions, while AI offers to supplement ideas or assistance. Thus, AI acts as a **Helper** to assist the user's creative process by offering structured guidance or brainstorming support (see Chompunuch & Lubart, 2025).

Specific Roles of AI in Human-Centered Collaboration

AI as Nanny

AI may facilitate the management of creative work by monitoring the creative process (setting deadlines and reminding human users to stay on track) and encouraging perseverance, which is a crucial need for creativity. Thus, AI, acting as a "Nanny," can facilitate the creative process by providing structure, support, and efficiency, ultimately enhancing human-centered collaboration. In short, an AI nanny provides tools to schedule and maintain creative activities (Lubart, 2005).

AI as Pen-pal

AI may facilitate communication of ideas. First, AI systems can now serve as notetakers in meetings. This reduces the load on human creators, allowing them to discuss ideas freely and benefit from meeting notes later. Second, AI systems may allow ideas to be posted in a community, tagged, and highlighted for further attention by others, through an AI-guided referral system. AI provides therefore advanced secretarial support.

AI as Coach

AI may facilitate the use of creativity enhancement techniques. AI, as embodied in expert systems knowledgeable in creativity-relevant techniques, can help human users implement creative thinking techniques, which can serve as analogs to jump-start the creative process. The user may engage in a certain cognitive process, such as divergent thinking, metaphorical thinking, or free association; AI can provide tutorials and exercises on how to enhance thinking (Lubart, 2005). So, the concept of *AI as a Coach* fits well for supporting the creative design systems. For example, the TRENDS system facilitates design tasks and couples a database of potential sources of inspiration for object design, which is based on experts' spontaneous sources and information on points of view to be considered, with a diagnosis-assistance module to guide designers to new ideas (Bonnardel & Bouchard, 2011). In short, AI coaches stimulate creative thinking by suggesting creative activity based on expert knowledge in the domain (Lubart, 2005).

Humans Remain the Primary Creators

Whereas AI provides structured guidance and support, such as proposing ideas, monitoring tasks, etc., the creative vision and final decisions remain with human creators. In addition, AI-generated creative outputs are shaped, reviewed, and integrated by the human creative lead. Thus, humans remain in full control or ownership over the project's direction and outcome.

Illustrations

Below are two illustrative cases demonstrating how AI can serve as a **Helper** in human-centered creativity. In each case, humans maintain full control while AI provides structured guidance and support.

YOLO Robot: Your Own Living Object

YOLO Robot (Fig. 8.2) is designed to stimulate creativity in children by promoting divergent and convergent thinking, which are essential for creative thought. It uses creativity techniques such as "contrast" and "mirror" to stimulate divergent and convergent thinking, respectively. These techniques involve providing stimuli that are either unrelated or related to the current creative task, encouraging creative exploration and reflection. The robot incorporates social behaviors informed by the Big Five Model of Personality, making it socially intelligent and engaging. This helps in creating an interactive environment without dominating the creative process. YOLO communicates using non-verbal expressive modalities such as lights and movements. This allows it to interact with children in a supportive manner, enhancing the play experience without overtaking it.

Research on the design process involved children at different creative stages, ensuring that the robot fits well within the children's play environment and supports their creativity without overshadowing it (Alves-Oliveira et al., 2019, 2020).

YOLO is used as a character in storytelling facilitation, providing stimuli for children to develop new storylines. This encourages children to lead the creative process while the robot supports their imagination. The robot's hardware and software are open-source, allowing customization to meet specific needs, which can further support creativity in a personalized manner (Alves-Oliveira et al., 2019, 2020). Additionally, studies have shown that interaction with social robots like YOLO can enhance human creativity by increasing engagement time and the number of creative ideas generated. This suggests that AI-based systems can play a significant role in supporting and expanding human creative potential (Lubart et al., 2021).

How It Works

- **User Input:** Children (or facilitators) introduce a creative task, such as storytelling, making art, or other imaginative play, and provide initial ideas or context. This shows that humans (children) remain the main drivers for the creative processes, setting the theme, storyline, and project goals.

Fig. 8.2 YOLO Robot (open access at https://www.wevolver.com/article/yolo-robot)

- **AI Assistance:** YOLO Robot analyses the child's actions and ideas and offers "contrast" prompts to light up new, novel, divergent ideas. Then, "mirror" prompts are used to help refine these ideas (convergent thinking). This socially intelligent behavior ensures a supportive and engaging interaction, primarily through non-verbal communication like lights and movements.
- **Feedback Loop:** As children respond to YOLO's prompts, the robot adjusts its suggestions, which maintains a balance between fostering creativity and giving the lead to children. This continuous give and take of ideas can help children explore, experiment, and refine their creations without overshadowing their creative input.

In sum, the YOLO Robot acts as a socially intelligent, creative helper who provides divergent and convergent thinking stimuli while maintaining a children-centered role in the creative process.

AI as a Helper, Using Specific GenAI Tools for Each Basic Creative Process Stage

Now we illustrate an example from Chompunuch and Lubart (2025). In this case, the researchers identify four basic creative processes: (1) Problem Identification and Framing, (2) Generating Ideas, (3) Evaluating Ideas, and (4) Deploying and Implementing Ideas. They mapped GenAI tools onto these stages of the creative process. Using an idea generation task about the United Nations SDG goal of Zero Hunger, Chompunuch and Lubart (2025) use **prompts** for specific GenAI tools for each stage of the creative process. For example,

- Problem Identification and Framing stage: Please frame the problem related to SDG Goal 2, Zero Hunger. We used **Research Problem Identifier AI** for this stage.
- Generating Ideas stage: Create many novel and innovative ideas for solving SDGs Global challenges such as Zero Hunger. (Your proposed idea should never be applied somewhere before.) We used **Ideamap AI** for this stage.
- Evaluating Ideas stage (Fig. 8.3): Select the best novel, innovative idea from the proposed ideas for solving Zero Hunger. Please give scores for each novel idea and why it is the best idea. We used **Sage Advisor AI** for this stage.

- Deploying and Implementing Idea stage (Fig. 8.4): Put the detail of the project outline; What is the concept of **Solar-Powered Mobile Grocery Shops**? Who is your target audience? What are the core features of your project? We used specific GenAI tools, such as **Idea-to-Prototyping AI, Leo AI, and Prototype App Generator,** for this stage.

In sum, the authors' proposition, "**AI as a Helper**," leverages AI's capabilities for data synthesis, novel idea generation, critique, and

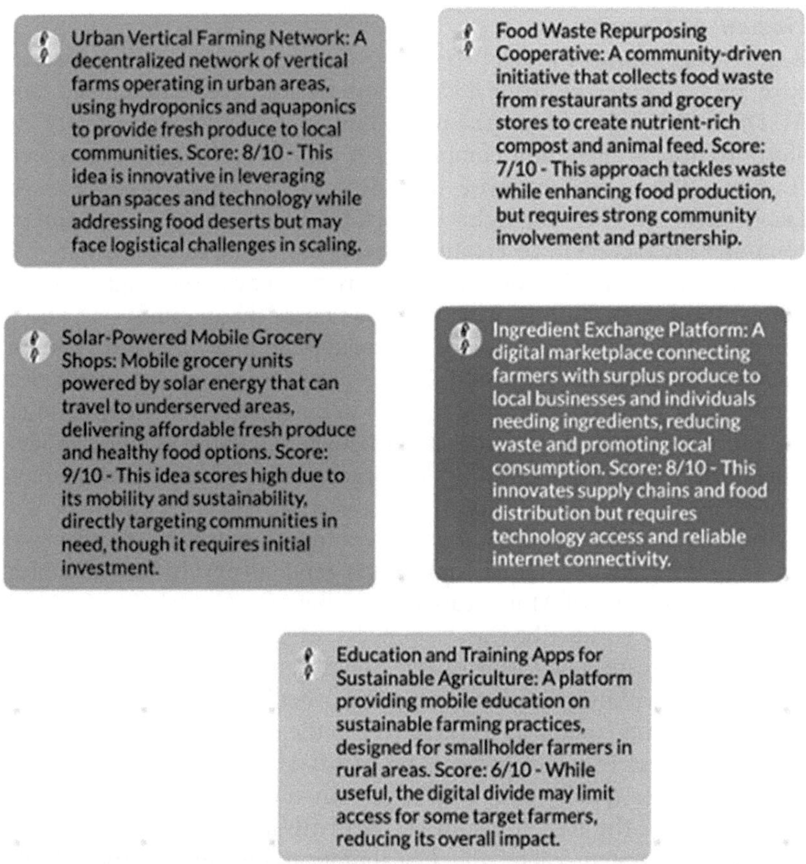

Fig. 8.3 Example results from idea evaluation

prototyping. By mapping GenAI tools to specific steps of the creative process and using AI features such as prompt-based text generation, image synthesis, etc., this framework/concept reduces ambiguity and shows when and how AI can add value. This approach ensures that human oversight remains central (human-centered), guiding AI outputs toward contextual awareness, ethical considerations, and high-quality solutions (Chompunuch & Lubart, 2025).

Mode 2

Human-AI Co-creation

Recent discussions in human-computer co-creativity show that the creative process emerges from a dense interaction between humans and AI. Davis (2013) introduces the term "human-computer co-creativity" to illustrate how humans and computers, as creative colleagues or partners (Lubart, 2005), can improvise in real-time to produce original work. Liapis et al. (2014) highlight that creativity emerging from human-computer interaction in co-creative systems cannot be solely attributed to either the human or the computer (AI), as it surpasses the original intentions of both contributors. Maher (2012) raises the question of "who is being creative" when humans and AI collaborate in a co-creative system. Building on this, Rezwana and Maher (2023) explore the interaction between collaborators (humans and AI) in co-creative systems, introducing the Co-Creative Framework for Interaction Design (COFI) model, which offers structured guidelines for organizing human-AI partnerships. Moreover, Vinchon et al. (2023) introduced a manifesto outlining scenarios for humans collaborating with machines during their creative tasks and suggesting a scenario that could guide generative AI toward an ethical and responsible use of AI in creativity, notably "Co-Cre-AI-tion."

Co-creation is described as "a process where humans and AI work together to create new forms of art, design, and innovation, with both contributing without one dominating the other" (Vinchon et al., 2023). Notably, humans could play a crucial role in the early and later parts of the creative process (e.g., identifying problems and refining AI-generated ideas to ensure quality and originality). In further studies, Kim et al. (2021) discuss the role of AI in co-creativity in design ideation. They introduced a system called the Collaboration Ideation Partner (CIP), which helps human users to generate design inspirations. The study

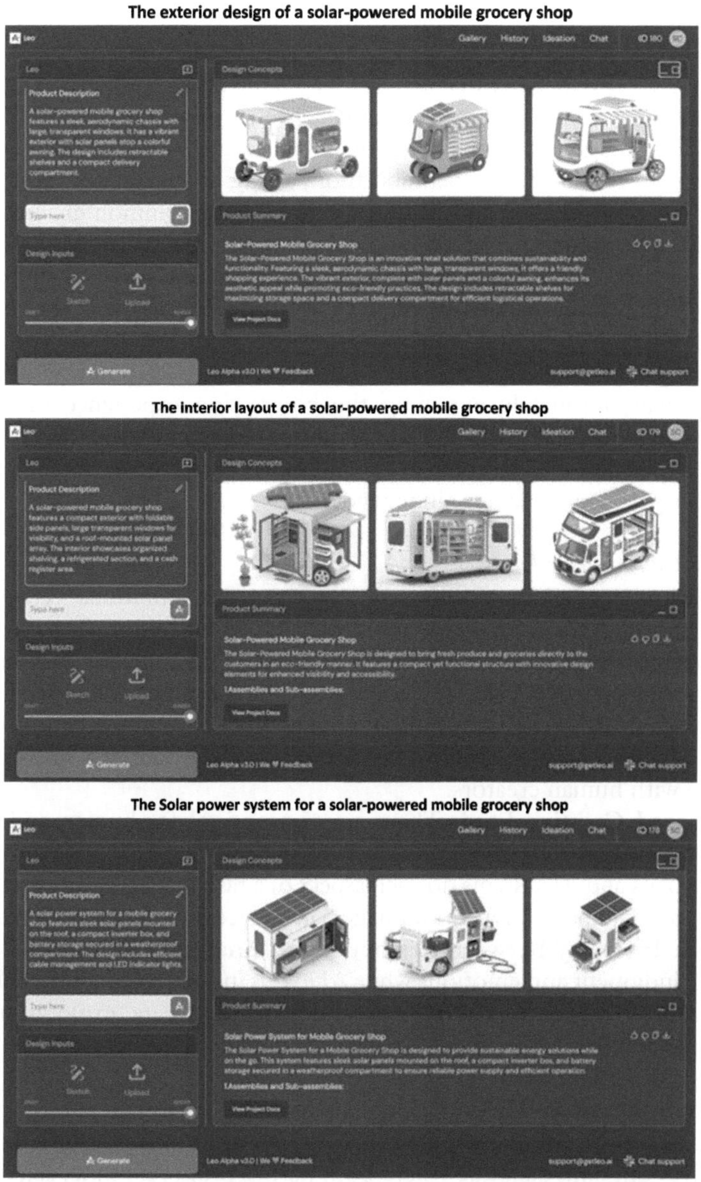

Fig. 8.4 Example results from deploying and implementing idea from Leo AI

highlights that AI can enhance human creativity and foster more novel and diverse ideas. Also, the participants perceived AI not as a mere tool but as a creative partner that actively shapes the design process by suggesting new solutions. Other co-creative systems in practice by Suh et al. (2021) highlight the significant role of AI in mediating social dynamics during collaborative music composition, acting as a facilitator rather than merely a tool. AI then contributes to establishing common ground, providing psychological safety, and promoting group progress, which enhances the creative process in teamwork. This presence of AI shifted the team's roles from creators to curators or producers, allowing the team to focus on evaluating and refining ideas. This positions AI as a potential "social glue," enabling smoother interactions and collaboration between individuals, ultimately enriching the co-creation experience in artistic endeavors (Suh et al., 2021). Overall, these studies emphasize that human-AI partnerships can produce novel and emergent creative outcomes that neither party could achieve independently.

AI as Colleague or Partner

According to Lubart (2005), one of the most ambitious versions of human-computer interaction for creativity is a genuine "partnership" between humans and computers where the machine operates not just as a mere tool but as a creative "colleague" or "partner." This perspective branches from research in artificial intelligence (AI), which demonstrates that computers can themselves be creative or contribute new ideas in dialogue with human creators.

Shared Creative Lead: There are two options for human and AI in terms of a shared creative lead. Humans do one aspect, and AI the other, or humans and AI systems do each aspect of a task in a back-and-forth way (e.g., both human and AI draw images) in a successive or co-active manner.

The Interplay Between Aesthetics and Analysis: Humans guide aesthetic judgment and emotions while AI excels in pattern generation, large-scale analysis, rapid prototyping, and expansive data exploration. By complementing each other's strengths, human-AI teams can iterate on ideas more efficiently.

Joint Decision-Making: Humans and AI collaboratively make key decisions by blending computational analysis with human insight. So, this involves ongoing dialogue where both parties propose, refine, select, and discard the creative ideas.

Illustrations

Below are two illustrative cases demonstrating how AI can serve as a **Colleague or Partner** in human-AI co-creation.

Shimon, the Robotic Musician

Shimon (Fig. 8.5) is an interactive robotic marimba player that is designed to improvise in real-time with a human pianist. It listens to human performance and builds upon it, creating a dynamic musical interaction (Weinberg & Driscoll, 2006). Shimon's technical design includes a musically and visually expressive motor-control system. It uses a gesture and anticipation-based improvisation system that emphasizes the importance of visual choreography in musical performance. The system models aspects of jazz improvisation, including chord coordination, tempo, and mutual inspiration between humans and robots. It allows melody freedom based on chord progressions and is structured into performance segments. Shimon builds on previous work with Haile, a robotic drummer, by expanding the instrumental range to a melodic instrument and increasing the motion range for more dynamic performances (Hoffman & Weinberg, 2010).

How It Works

1. **Real-Time Listening and Analysis:** Shimon "listens" to a human musician's performance by capturing pitch, rhythm, chord progressions, and tempo. Then, the system identifies key musical elements (chords, beats) to understand the current improvisational context.
2. **Improvisation and Interaction:** Shimon uses predictive models to guess what the human might play next, adapting it to melody or rhythm, which responds in real-time. Like a Jazz Musician, Shimon exchanges musical ideas to create a dynamic "call-and-response" interplay.
3. **Robotic Performance and Visual Expression:** Shimon's arms or mallets strike the marimba in various positions to produce melodic lines and rhythmic patterns. Shimon's head and body movements are designed to visibly match the musical flow and enhance the sense of a live co-performing partner.

Fig. 8.5 Shimon Robot (open access at https://www.shimonrobot.com/)

Rootport's "Cyberpunk: Peach John"

Rootport's **"Cyberpunk: Peach John" Manga** series (Fig. 8.6) was created using MidJourney, an AI image generator known for producing highly stylized and artistic visuals. It demonstrates a harmonious blend of artistic vision and AI technology (MidJourney).

How It Works

- **Story Concept Development (Human-Led):** Rootport (the human author) starts with a strong creative vision for his series, including themes, characters, and story settings. This concept reflects Rootport's artistic initiative, knowledge, and expertise that AI cannot independently create.
- **Prompt Writing (Human Inputs):** Rootport communicates his creative vision to MidJourney AI through prompts. These prompts describe key elements of his artwork, such as "*Cyberpunk city at night, glowing neon signs, Japanese temple in the background, fog and*

Fig. 8.6 Rootport "Cyberpunk: Peach John" Manga series (from Midjourney)

rain for a moody atmosphere." He refines and iterates prompts and modifies details based on how AI interprets them.

- **Image Generation (AI-Led):** MidJourney uses the prompts to generate different artistic outputs. Each output is a unique visual interpretation. It offers a blending of the elements of cyberpunk and Japanese culture that are described by Rootport.
- **Curation and Refinement (Human-AI Collaboration):** Rootport evaluates the AI-generated images and selects those that align with his creative vision. If needed, Rootport refines his prompts and re-generates images until they get closer to his desired look.
- **Story Integration (Human-Led):** Rootport uses refined images to craft a narrative. In "Cyberpunk: Peach John," the story explores themes of isolation, survival, and the clash between tradition and modernity.
- **Iterative Improvement (Human-AI Collaboration):** As Human-AI co-creation is an iterative process, Rootport may contrib-

ute to experimenting with new prompts and images, gathering feedback from the audiences, and adapting the series' direction based on creative insight provided by AI outputs (Holland et al., 2023).

Mode 3

AI as Principal Creator (AI-Centered)

In this mode, we focus on AI- or robot-centered approaches (Dautenhahn, 2007; Lubart et al., 2021). We emphasize AI autonomous systems that are designed to produce work that is intended to be original and potentially valuable (Alves-Oliveira et al., 2020; Pan et al., 2010; Peng et al., 2015) or fully autonomous systems that "are built to generate creative artifacts that later are judged by humans to be creative" (Colton, 2012; Colton et al., 2015; Elgammal, 2017; Mazzone & Elgammal, 2019). Our mode is illustrated with the highest AI involvement and minimal human intervention in the creative process (Zhang et al., 2023).

In several cases, they showcase various robots, some with humanoid appearances (Kato et al., 1987; Calinon et al., 2005; Gazeau & Zeghloul, 2012; Tresset & Leymarie, 2013; Luo & Liu, 2018) and create meaningful new output in complex tasks. The well-known examples we illustrate in this section are AARON, an autonomous artist by Cohen (1995), and AlphaGo from Google DeepMind. These two examples highlight the adaptability of **AI-centered creativity** across different fields (art and game strategy).

AI as Creative Agents

Advanced AI systems and now Creative Agents (such as social robots) can independently generate realistic images and videos, compose music, write creative content, and design objects. Several key approaches and techniques were used behind these advances, notably generative models (e.g., Generative Adversarial Networks (GANs), diffusion models, Transformer Models), reinforcement learning (RL) agents, and hybrid models. We overview these approaches and illustrate how they contribute to independently creating artwork.

Generative models such as GANs (Goodfellow et al., 2014), variational autoencoders (VAEs) (Kingma & Welling, 2013), diffusion models (Ho et al., 2020), and transformer-based architectures (Vaswani et al., 2017) are the ones that are the most widespread as of today. These models utilize adversarial training in latent space to synthesize realistic images, compelling musical sequences, and coherent narratives. Thus, their ability to mimic human creative outputs is reshaping creative industries by fostering innovation in digital content production, enhancing new creative collaboration workflows between humans and AIs, and redefining our understanding of artistic expression.

Reinforcement learning (RL) agents contribute to the creative domain by learning to make decisions by interacting with an environment and iterative feedback in the form of rewards (Sutton & Barto, 2018). This model can be adapted to creative tasks where the "environment" might include artistic constraints or audience feedback. For example, in music composition systems, interactive art, and dynamic game design, RL agents make sequential decisions and use trial-and-error processes, which sometimes incorporate human feedback to help fine-tune outputs that meet specific guidelines, balancing creativity with coherence and safety (Christiano et al., 2017). The pioneer example that uses reinforcement learning techniques was AlphaGo.

Lastly, hybrid models, combining generative models and RL, can produce a diverse range of creative outputs.

Overall, these models have transformed and reshaped creative disciplines by employing data-driven learning to generate artistic outputs. They not only mimic or replicate human creativity through techniques like adversarial training and latent space manipulation (Goodfellow et al., 2014; Kingma & Welling, 2013; Ho et al., 2020; Vaswani et al., 2017) but also open new possibilities for innovation in digital art, music, and narrative writing.

Human Guidance and Feedback

Although AI systems lead as creators in this third category, human contribution remains essential. In Fig. 8.7, we present the "AI Art Creative Process Diagram" (Elgammal, 2018), which highlights **human intervention in pre-curation, tweaking, and post-curation that are critical** to achieving results that align with artistic standards and expectations.

Mazzone and Elgammal (2019) discuss the important role of humans that remains crucial during the post-curation process (Fig. 8.7). They review numerous images created by AI and select the best ones that meet their artistic standards. This **post-curation** step is essential because it allows the artists to improve the AI-generated images to match their creative vision and appeal to viewers. The post-editing or post-curation process demonstrates the collaboration between human creativity and AI-generated art. The artists work to enhance and finalize the AI's output, ensuring that the artwork is of top quality and resonates with the audience (Mazzone & Elgammal, 2019). Therefore, humans play a central role in guiding the AI creative process (even human-minimal intervention) by providing objectives, constraints, and aesthetic judgments that shape the desired outcome.

Illustrations

Below are two illustrative cases demonstrating how AI can serve as a **Principal Creator** in AI-centered creativity. In each case, AI takes the lead in the creative process while humans provide structured guidance and feedback.

AARON by Harold Cohen

AARON is an autonomous AI system designed to generate original artworks (Fig. 8.8). This is an AI pioneer system that functions as a principal creator. With the highest AI involvement, we can see AI leading the creative process. But how does AARON function as a Principal Creator (AI-centered) in Mode 3?

First, Cohen encoded **knowledge and rules** for color theory, form, and constraints that reflect how shapes and figures could be represented in

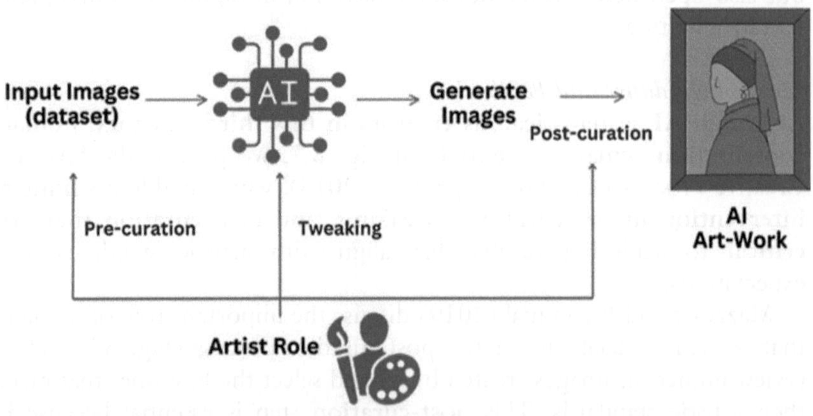

Fig. 8.7 AI-art creative process diagram (adapted from Elgammal, 2018)

a coherent scene. Further, AARON uses these rules to **autonomously generate** paintings or drawings. AARON decides how to shape figures, where to place lines, and how to color the canvas. In this sense, Harold Cohen **supervised** the system's development and made iterative improvements to AARON. Once trained, AARON made creative decisions on its own with minimal human intervention.

Some AARON versions were executed and collaborated with robot painter arms to physically produce the artworks on canvas.

As shown in AARON, we can illustrate their work steps by referring to Fig. 8.7, an **AI-art creative process diagram** (Elgammal, 2018).

1. **Pre-curation (Setup and Rule-Based)**
   Harold Cohen developed AARON by encoding artistic knowledge and rules related to (1) shape and composition, (2) proposition and perspective, and (3) color theory. Thus, Cohen provided high-level creative guidelines to AARON.

Fig. 8.8 The Robotic artist AARON in Living Color (https://whitney.org/exhibitions/harold-cohen-aaron)

2. **AI Leading the Process (Autonomous)**

   AARON then autonomously decides what to draw based on the rule set and generates output each time by creating Abstract forms and Figurative works (e.g., human figures, plants, or animals).

   AARON enhances the creative process by "Sketching" and layering elements like a human artist and applying composition and color choices according to its artistic logic.
3. **Post-curation (Setup and Rule-Based)**

   After AARON generates artwork, a **Human Curator** (Cohen or an art gallery) could **Select** artworks for exhibition and **Refine** or **Interpret** the meaning of AARON's pieces in the context of art shows. So, while AARON makes creative decisions, humans also provide **Context** for displaying and publishing its works. In addition, Cohen provides feedback and updates or expands AARON's rules over time (Cohen, 1995).

AlphaGo by Google DeepMind

AlphaGo is an AI board game system that is trained using a combination of supervised learning (based on human game data) and reinforcement learning (self-learning). AlphaGo generated innovation strategies that defeated the world's top Go players (Fig. 8.9). A well-known example is Move 37 during the match against Lee Sedol, which was considered unconventional yet brilliant. AlphaGo showcases its ability to create new strategic ideas beyond human intuition. Next, we demonstrate how it works as a Principal Creator (AI-centered) in Mode 3.

How It Works

1. **Pre-curation:** Humans set up initial training by providing game data and establishing learning objectives.
2. **Tweaking:** Developers observe gameplay performance and adjust training or algorithms if needed.
3. **Post-curation:** Professional Go players and analysts study AlphaGo strategies to incorporate its creative insights into their gameplay, advancing human knowledge of the game.

This example demonstrates AI-centered creativity by taking the lead in the creative process, generating novel strategies, and reshaping the human

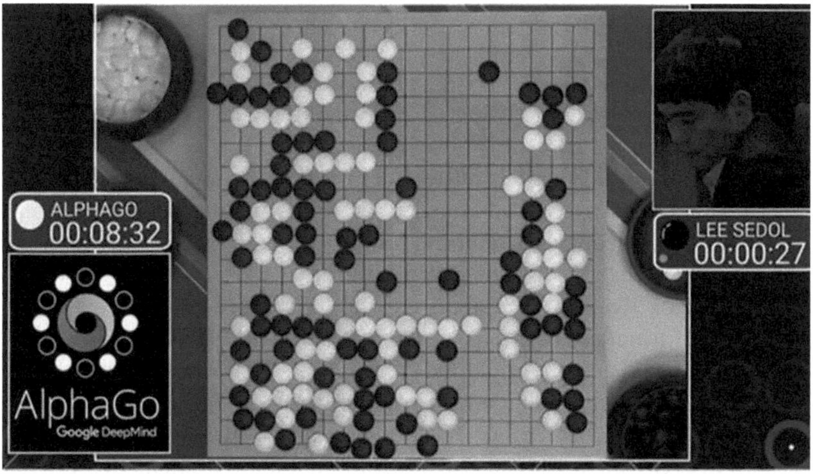

Fig. 8.9 AlphaGo, AI board game Go match against Lee Sedol (open access from https://deepmind.google/research/projects/alphago/)

understanding of complex games (Silver et al., 2016; DeepMind, 2016; Sadler & Regan, 2019).

Ethics and Issues of Using AI

As these AI-generated techniques evolve, they challenge traditional notions of creativity and authorship while also promoting collaborative workflows between humans and machines. This ongoing development is likely to expand the scope of interdisciplinary research, raise important ethical questions, and drive innovation. However, such developments also require careful examination of issues such as copyright, ownership, artistic control, and bias in AI-generated content. Moving forward and balancing the potential benefits of AI automation, we need transparent practices and responsible governance to preserve the trust, accessibility, and integrity of the creative process. The issues, such as copyright, will be clearly impacted by the type of human-AI collaboration given the three modes of collaboration highlighted in our presentation.

Hybrid Intelligence and Beyond

The concept of "Hybrid Intelligence" (HI) refers to "the ability to achieve complex goals by combining human and artificial intelligence (Fig. 8.10), thereby reaching superior results to those each of them could have accomplished separately, and continuously improve by learning from each other" (Dellermann et al., 2019).

There are three core concepts of Hybrid Intelligence/Creativity:

- **Collectively Hybrid Intelligence**: In the HI system, Tasks are performed collectively; consequently, activities conducted by each agent are conditionally dependent. This means that while the agents work together, their goals may not always align.
- **Superior Results**: The socio-technical system of HI achieves a level of performance in specific tasks that surpasses what any individual agent could accomplish alone. This is a key characteristic that distinguishes hybrid intelligence from other forms of AI applications.
- **Continuous Learning**: A central aspect of HI is the continuous improvement of the system. Both human and AI agents learn from each other, enhancing their performance over time. This learning process is crucial for the development and effectiveness of the system (Dellermann et al., 2019; Rafner et al., 2022).

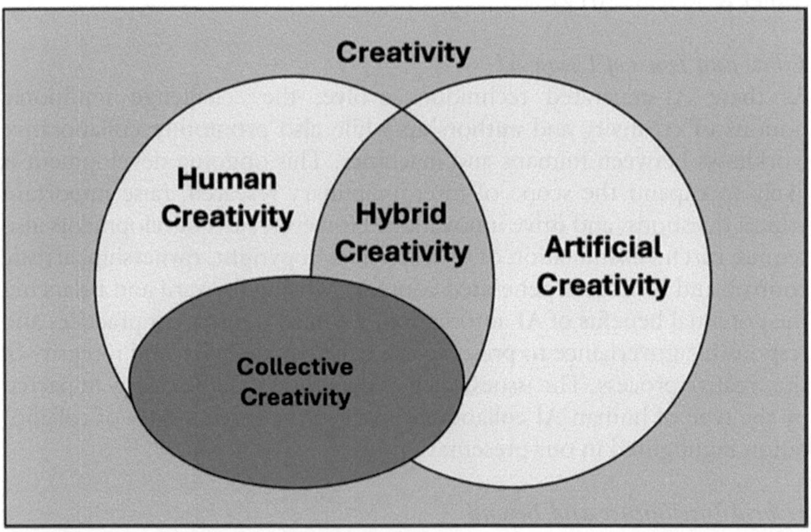

Fig. 8.10 Conceptual integration of hybrid intelligence/creativity (from Dellermann et al., 2019)

The Hybrid Intelligence (HI) concept fits well with our three modes of human-AI collaboration: **AI as Helper, AI as Partner, and AI as Principal Creator**, and offers an overall theoretical concept that unites the three modes of collaboration that we have explored.

Conclusion

In this chapter, we provided an overview of three modes of human-AI collaboration. We also outlined the various roles that AI can play in the creative process. These modes demonstrate how AI can serve as a **Helper, Partner, or Principal Creator** that offers unique opportunities and challenges.

For **Mode 1: Human-Centered Creativity (AI as Helper)**, we discussed how AI supports human creativity by providing creative suggestions and automating some ideation tasks. We illustrated examples like YOLO Robot and the Case of combining GenAI tools for all stages of the creative process.

For **Mode 2: Human-AI Co-creation (AI as Partner)**, we discussed the collaboration between humans and AI and the role of humans as colleagues or partners in co-creation. We also illustrated this with Shimon, the robotic musician who partners with humans in music composition. In addition, the example of Rootport shows how he co-created with MidJourney AI to produce highly stylized and artistic visuals for his Manga series "Cyberpunk: Peach John."

Finally, for **Mode 3: AI-Centered Creativity (AI as Principal Creator)**, we provided examples such as AARON, an autonomous artist, and AlphaGo, which demonstrated strategic creativity, where AI takes the lead with minimal human intervention.

Overall, these examples illustrate the complementary roles that humans and AI play in the creative process. In each mode, AI enhances human creativity through support, co-creation, or autonomous creation, while humans remain central to guidance, evaluation, and improvement. In the AI as Helper mode, humans maintain full control and lead the creative process. In the AI as Partner mode, both humans and AI co-create and balance their contributions more or less symmetrically. In the AI as Principal Creator mode, AI takes the lead in the creative process, but human curation of the work ensures it achieves its artistic goals. These modes demonstrate the evolving relationship between human and AI creativity, emphasizing how successful collaborations balance human intuition and expertise with AI-driven opportunities.

References

Alves-Oliveira, P., Arriaga, P., Paiva, A., & Hoffman, G. (2019). Guide to build YOLO, a creativity-stimulating robot for children. *HardwareX, 6*, e00074.

Alves-Oliveira, P., Arriaga, P., Paiva, A., & Hoffman, G. (2020, March). Yolo-your own living object. In *Companion of the 2020 ACM/IEEE International Conference on Human-Robot Interaction* (pp. 638–638). ACM/IEEE.

Bonnardel, N., & Bouchard, C. (2011, November). Towards supporting creative design: Analysis of the use of the TRENDS system according to designers' expertise. In *Proceedings of the 8th ACM Conference on Creativity and Cognition* (pp. 315–316). ACM.

Calinon, S., Guenter, F., & Billard, A. (2005, April). Goal-directed imitation in a humanoid robot. In *Proceedings of the 2005 IEEE International Conference on Robotics and Automation* (pp. 299–304). IEEE.

Chompunuch, S., & Lubart, T. (2025). AI as a helper: Leveraging generative AI tools across common parts of the creative process. *Journal of Intelligence, 13*(5), 57.

Christiano, P. F., Leike, J., Brown, T., Martic, M., Legg, S., & Amodei, D. (2017). Deep reinforcement learning from human preferences. *Advances in Neural Information Processing Systems, 30*.

Cohen, H. (1995). The further exploits of AARON, painter. *Stanford Humanities Review, 4*(2).

Colton, S. (2012). The painting fool: Stories from building an automated painter. In *Computers and creativity* (pp. 3–38). Springer.

Colton, S., Halskov, J., Ventura, D., Gouldstone, I., Cook, M., & Ferrer, B. P. (2015, June). The painting fool sees! New projects with the automated painter. In *Proceedings of the International Conference on Computational Creativity* (pp. 189–196).

Dautenhahn, K. (2007). Socially intelligent robots: Dimensions of human–robot interaction. *Philosophical Transactions of the Royal Society, B: Biological Sciences, 362*(1480), 679–704.

Davis, N. (2013). Human-computer co-creativity: Blending human and computational creativity. In *Proceedings of the AAAI conference on artificial intelligence and interactive digital entertainment* (Vol. 9, pp. 9–12).

Davis, N., Hsiao, C. P., Popova, Y., & Magerko, B. (2015). An enactive model of creativity for computational collaboration and co-creation. *Creativity in the Digital Age*, 109–133.

DeepMind. (2016). *AlphaGo: The story so far*. DeepMind Blog.

Dellermann, D., Ebel, P., Söllner, M., & Leimeister, J. M. (2019). Hybrid intelligence. *Business & Information Systems Engineering, 61*(5), 637–643.

Doncieux, S., Chatila, R., Straube, S., & Kirchner, F. (2022). Human-centered AI and robotics. *AI Perspectives, 4*(1), 1.

Elgammal, A. (2017). *Can: Creative adversarial networks, generating "art" by learning about styles and deviating from style norms.* arXiv preprint arXiv:1706.07068, 6, 2017.

Elgammal, A. (2018). *A portrait created by Artificial Intelligence has blurred the line between machine and artist.* Scroll.In. https://scroll.in/article/898493/a-portraitcreated-by-artificial-intelligence-has-blurred-the-line-between-machine-and-artist

Gazeau, J. P., & Zeghloul, S. (2012). The artist robot: A robot drawing like a human artist. In *2012 IEEE International Conference on Industrial Technology* (pp. 486–491). IEEE.

Glaveanu, V. P., Hanchett Hanson, M., Baer, J., Barbot, B., Clapp, E. P., Corazza, G. E., et al. (2020). Advancing creativity theory and research: A socio-cultural manifesto. *The Journal of Creative Behavior, 54*(3), 741–745.

Goodfellow, I., Pouget-Abadie, J., Mirza, M., Xu, B., Warde-Farley, D., Ozair, S., et al. (2014). Generative adversarial nets. *Advances in Neural Information Processing Systems, 27.*

Ho, J., Jain, A., & Abbeel, P. (2020). Denoising diffusion probabilistic models. *Advances in Neural Information Processing Systems, 33.*

Hoffman, G., & Weinberg, G. (2010). Shimon: An interactive improvisational robotic marimba player. In *CHI'10 extended abstracts on Human Factors in Computing Systems* (pp. 3097–3102).

Holland, O., Sugiura, N., & Jozuka, E. (2023). This is Japan's first AI-generated manga comic. But is it art? *CNN.* https://www.cnn.com/style/article/japan-first-ai-generated-manga-art-intl-hnk

Karimi, P., Grace, K., Maher, M. L., & Davis, N. (2018). *Evaluating creativity in computational co-creative systems.* arXiv preprint arXiv:1807.09886.

Kato, I., Ohteru, S., Shirai, K., Matsushima, T., Narita, S., Sugano, S., et al. (1987). The robot musician 'wabot-2'(waseda robot-2). *Robotics, 3*(2), 143–155.

Kim, J., Maher, M. L., & Siddiqui, S. (2021). Collaborative ideation partner: Design ideation in human-AI co-creativity. In *CHIRA* (pp. 123–130).

Kingma, D. P., & Welling, M. (2013). *Auto-encoding variational Bayes.* arXiv preprint arXiv:1312.6114.

Latour, B. (2007). *Reassembling the social: An introduction to actor-network-theory.* OUP.

Liapis, A., Yannakakis, G. N., & Togelius, J. (2014). *Computational Game Creativity.* ICCC.

Lubart, T. (2005). How can computers be partners in the creative process: Classification and commentary on the special issue. *International Journal of Human-Computer Studies, 63*(4–5), 365–369.

Lubart, T., Esposito, D., Gubenko, A., & Houssemand, C. (2021). Creativity in humans, robots, humbots. *Creativity. Theories–Research-Applications, 8*(1), 23–37.
Luo, R. C., & Liu, Y. J. (2018). Robot artist performs cartoon style facial portrait painting. In *2018 IEEE/RSJ International Conference on Intelligent Robots and Systems* (pp. 7683–7688). IEEE.
Maher, M. L. (2012, May). Computational and collective creativity: Who's being creative? In *ICCC* (pp. 67–71).
Mao, Y., Rafner, J., Wang, Y., & Sherson, J. (2024). HI-TAM, a hybrid intelligence framework for training and adoption of generative design assistants. *Frontiers in Computer Science, 6*, 1460381.
Mazzone, M., & Elgammal, A. (2019, February). Art, creativity, and the potential of artificial intelligence. In *Arts* (Vol. 8, No. 1, p. 26). MDPI.
Oh, C., Song, J., Choi, J., Kim, S., Lee, S., & Suh, B. (2018, April). I lead, you help but only with enough details: Understanding user experience of co-creation with artificial intelligence. In *Proceedings of the 2018 CHI Conference on Human Factors in Computing Systems* (pp. 1–13).
Pan, Y., Kim, M. G., & Suzuki, K. (2010). A robot musician interacting with a human partner through initiative exchange. In *Proceedings of the Conference on New Interfaces for Musical Expression* (pp. 166–169).
Peng, H., Zhou, C., Hu, H., Chao, F., & Li, J. (2015). Robotic dance in social robotics—A taxonomy. *IEEE Transactions on Human-Machine Systems, 45*(3), 281–293.
Rafner, J., Dellermann, D., Hjorth, A., Verasztó, D., Kampf, C., Mackay, W., & Sherson, J. (2022). Deskilling, upskilling, and reskilling: A case for hybrid intelligence. *Morals & Machines, 1*(2), 24–39.
Rezwana, J., & Maher, M. L. (2023). Designing creative AI partners with COFI: A framework for modeling interaction in human-AI co-creative systems. *ACM Transactions on Computer-Human Interaction, 30*(5), 1–28.
Sadler, M., & Regan, N. (2019). *Game changer: AlphaZero's groundbreaking chess strategies and the promise of AI.* New in Chess.
Shneiderman, B. (2002). Creativity support tools. *Communications of the ACM, 45*(10), 116–120.
Shneiderman, B. (2007). Creativity support tools: Accelerating discovery and innovation. *Communications of the ACM, 50*(12), 20–32.
Silver, D., et al. (2016). Mastering the game of Go with deep neural networks and tree search. *Nature, 529*(7587), 484–489.
Suh, M., Youngblom, E., Terry, M., & Cai, C. J. (2021, May). AI as social glue: Uncovering the roles of deep generative AI during social music composition. In *Proceedings of the 2021 CHI Conference on Human Factors in Computing Systems* (pp. 1–11).
Sutton, R. S., & Barto, A. G. (2018). *Reinforcement Learning: An Introduction* (2nd ed.). MIT Press.

Tresset, P., & Leymarie, F. F. (2013). Portrait drawing by Paul the robot. *Computers & Graphics, 37*(5), 348–363.

Vaswani, A., Shazeer, N., Parmar, N., Uszkoreit, J., Jones, L., Gomez, A. N., et al. (2017). Attention is all you need. *Advances in Neural Information Processing Systems, 30.*

Vinchon, F., Lubart, T., Bartolotta, S., Gironnay, V., Botella, M., Bourgeois-Bougrine, S., Burkhardt, J.-M., Bonnardel, N., Corazza, G.E., Glăveanu, V., Hanchett Hanson, M., Ivcevic, Z., Karwowski, M., Kaufman, J.C., Okada, T., Reiter-Palmon, R. & Gaggioli, A. (2023). Artificial intelligence & creativity: A manifesto for collaboration. *The Journal of Creative Behavior, 57*(4), 472–484.

Weinberg, G., & Driscoll, S. (2006). Toward robotic musicianship. *Computer Music Journal*, 28–45.

Wingström, R., Hautala, J., & Lundman, R. (2024). Redefining creativity in the era of AI? Perspectives of computer scientists and new media artists. *Creativity Research Journal, 36*(2), 177–193.

Yannakakis, G. N., Liapis, A., & Alexopoulos, C. (2014). Mixed-initiative co-creativity. In *International Conference on Foundations of Digital Games.*

Zhang, M., Cheng, Z., Shiu, S. T. R., Liang, J., Fang, C., Ma, Z., et al. (2023). Towards human-centred AI-co-creation: A three-level framework for effective collaboration between human and AI. In *Companion publication of the 2023 Conference on Computer Supported Cooperative Work and Social Computing* (pp. 312–316).

Zhang, R., McNeese, N. J., Freeman, G., & Musick, G. (2021). "An ideal human" expectations of AI teammates in human-AI teaming. *Proceedings of the ACM on Human-Computer Interaction, 4*(CSCW3), 1–25.

CHAPTER 9

From Social Encounters to Sociotechnical Creative Collaboration: Perceiving and Engaging with Robots

Nathalie Bonnardel, Maejda Marcon, Ladislav Moták, and Clément Belletier

Introduction

In a rapidly changing world, where technological, economic, and societal transformations are occurring at a rapid pace, individuals are evolving in increasingly complex, unstable, and often unpredictable environments.

N. Bonnardel
Aix-Marseille Université, Centre de recherche en Psychology de la Connaissance, du Langage et de l'Emotion (PSYCLE, UR 3273), Aix-en-Provence, France

Institute on Creativity and Innovations of Aix-Marseille (InCIAM), Aix-en-Provence, France
e-mail: nathalie.bonnardel@univ-amu.fr

M. Marcon
Aix-Marseille Université, Centre de recherche en Psychology de la Connaissance, du Langage et de l'Emotion (PSYCLE, UR 3273), Aix-en-Provence, France

Institute on Creativity and Innovations of Aix-Marseille (InCIAM), Aix-en-Provence, France

Université Clermont Auvergne, Laboratoire de Psychologie Sociale et Cognitive (LAPSCO), CNRS, Clermont-Ferrand, France
e-mail: maejda.marcon@univ-amu.fr

N. Bonnardel (ed.), *Collaborations in Creativity*, The Seven C's of Creativity, https://doi.org/10.1007/978-3-032-07753-0_9

Thus, it is crucial that individuals be able to mobilize both cognitive and psychosocial capabilities, and more and more frequently technical skills, that enable them to adapt effectively to new situations. More specifically, the capability to be creative, to deal with the unexpected with flexibility, and to solve complex problems without relying on standardized procedures is a fundamental requirement for meeting contemporary challenges in various areas (see, for instance, Bonnardel et al., 2023; Lubart, 2018). These abilities are no longer reserved for exceptional contexts or certain professions, but are now becoming essential cross-cutting skills in many fields of activity and must therefore be taken into account when developing curricula or training programs (e.g., Vincent-Lancrin et al., 2019; Bonnardel & Didier, 2016, 2020). In addition to contexts explicitly geared toward creative production, creativity also plays an essential role in dealing with ill-defined situations that challenge the status quo or hinder progress. Moreover, as many complex creative situations require knowledge that is distributed among several individuals rather than held by a single person, collective creative activities are becoming increasingly common in our society, hence the interest of this book.

Traditionally, *collaboration* is understood as a complex process in which humans with intentionality, efficient communication skills, and a shared understanding and awareness of the situation, pursue a common goal and work together to achieve that goal. Thus, as described in Chap. 1 (by Bonnardel et al.) and in Chap. 2 (by Baker et al.), collaboration in creativity is usually based on an integrated form of interaction, in which individuals not only contribute regarding their field of expertise, but work together to produce something new and possibly adapted or contextually

L. Moták
Université Clermont Auvergne, Laboratoire de Psychologie Sociale et Cognitive (LAPSCO), CNRS, Clermont-Ferrand, France
e-mail: ladislav.motak@uca.fr

C. Belletier (✉)
Université Clermont Auvergne, Laboratoire de Psychologie Sociale et Cognitive (LAPSCO), CNRS, Clermont-Ferrand, France

LAPSCO CNRS & Institut Universitaire de France (IUF), Clermont-Ferrand, France
e-mail: clement.belletier@uca.fr

appropriate—in line with definitions of creativity—in pursuit of a shared objective. It requires that collaborators adhere to a shared vision of what they are trying to achieve (e.g., the creative task to be performed), which contributes to aligning their motivations, actions, and collective efforts, as highlighted when applying the COFOR model to creative activities (see Bonnardel, 2024, and Chapter 1). The interactions that underpin human collaboration are therefore dependent on both individual characteristics (cognitive abilities, personality, intrinsic motivation, knowledge), group characteristics (norms, cohesion, size, roles, tasks, diversity, problem-solving techniques), and organizational characteristics (culture, resources, rewards, strategy, structure, technology) as highlighted in the interactionist model proposed by Woodman et al. (1993) and described in Chap. 3 (by Girandola and Bonnardel).

Moreover, collective creative activities require an implicit understanding of the context, continuous mutual adjustments, and, sometimes, even empathy toward future users as it is the case in prospective design contexts (e.g., Bonnardel & Pichot, 2020; Brangier et al., 2011; Robert & Brangier, 2012). Dimensions related to these processes—such as "sustaining mutual understanding," "information exchanges for problem solving," and "argumentation and reaching consensus"—are especially taken into account and associated with specific indicators in the collaboration quality assessment grid proposed by Détienne et al. (2012a, 2012b) and described in Chap. 2 (by Baker et al.). The cognitive and interactional processes as well as behaviors developed during team problem-solving activities have also been taken into account in the analysis and coding system, called SPARC (Social Process Assessment for Recognizing Creativity), presented in Chap. 4 (by Christenson et al.).

In addition to the conceptual and methodological contributions provided in this book, various types of collective creative situations based on interhuman collaboration have been described in depth. For instance, collective design situations have been analyzed both in Chap. 1 (by Bonnardel et al.) and in Chap. 2 (by Baker et al.). In particular, several studies presented in Chap. 1 were related to prospective ergonomics and aimed at designing products and services for the future. As described in the first chapter, different types of participants were asked to design either a new kind of apartment (a connected apartment) or a bus adapted to the needs of seniors and their caregivers, while anticipating the future cognitive and physical limitations of seniors. Other creative tasks, also forward-looking, aimed to imagine ways of reducing traffic jams and the occurrence of road accidents, or new uses, objects, or forms of communication in a distant future (in 2050). The authors especially analyzed whether different methods (e.g., de Bono's six thinking hats, design fiction, dynamic personas) associated with experimental conditions can foster both creativity and collaboration while taking into consideration future users' needs. Three other chapters focused on real-life collective creative situations in the field. Thus, a systemic approach to scientific creativity has been presented in Chap. 5 (by Yamanouchi et al.), which refocused analysis on the collaborative dynamics between idea producers and the field that evaluates them in the context of peer review. The authors proposed to go beyond the traditional opposition between individual generation and collective evaluation, broadening the notion of the "field" of evaluation to include interdisciplinary and societal actors, in line with the evolution of science-society relations and the principles of open science. In Chap. 6 (by Bourgeois-Bougrine & Le Bos), an original approach of embodied cognition was proposed and applied to analyze the writing of stories by screenwriters in the film industry. According to this approach, human cognition is not solely in the head but is a result of dynamic interactions between the brain, the body, and the world. It allowed the authors to discuss the creative collaboration based on the "4E" perspective (embodied, embedded, extended, and enactive). A physical and virtual narrative space was also described, in which professional writers can enter at the same time, explore, experiment, and experience all the possibilities offered by a fictional world in order to produce compelling narratives that can serve as a means of reflecting on today's societal challenges and imagining alternative futures. Another real-life situation has been analyzed in Chap. 7 (by Nakakogi et al.) in a museum-like environment, where historical statistical tables can serve as tools for citizen inquiry and participatory knowledge creation. Inspired by the principles of museum design, based on experiential interaction, peer influence, and cumulative expression, two tools have been developed to enable collaboration distributed across time and space: one allowing collaborative annotation workshops, and the other one providing participants with a digital platform for exploring these annotations asynchronously. The authors described how participants collectively construct meaning from data, illustrating a form of situated and personal creativity (mini-c), and highlighted the potential of archived data as participatory cultural resources that foster shared reflection and collective creativity. Therefore, various forms of collaboration in creative activities have been described and analyzed in depth, allowing new aspects of creative and collaborative activities to be explored.

Another central aspect of creative and collaborative activities, which is becoming increasingly relevant in our society, is also discussed in this book: *the transition from human-human collaboration to human-system collaboration, such as human-AI or human-social robot collaboration.* It has been explored in the previous chapter (by Chompenuch et al.) and is further developed in this chapter. Human-system collaboration disrupts the "classic" framework and raises several questions, such as the following: Can we still talk about "collaboration" in the strong sense of the term when one of the partners is an artificial entity, often limited in cognitive autonomy, contextual interpretation, and initiative? This shift requires us to rethink the foundations of collaboration. While creativity has long been considered a cognitive and social process embedded in human collectives, the rise of computational agents, whether algorithmic or embodied, has introduced new dynamics and promoted more and more human-computer and/or human-robot interactions. Therefore, this concluding chapter proposes reflections and perspectives about what it means to collaborate creatively in a world where humans, AI, and robots increasingly co-create. Indeed, the evolution of collaborative practices is undergoing a profound transformation since collaboration is nowadays increasingly mediated, supported, or enacted by artificial systems, ranging from software agents to social robots. Indeed, in various domains, humans increasingly interact with systems to create or co-create with a generative artificial intelligence (see Corazza, 2025), and sometimes the AI is embodied in a social robot, as described in the previous chapter and in this one. Especially, two fundamental questions about collaboration with a social robot are addressed in this chapter:

- How does an encounter with a social robot constitute a social encounter?
- Which human-robot interactions elicit increased creativity, and which interactions do not work? And why?

Therefore, we examine how users' perceptions and engagement strategies adapt to new forms of artificial agency in order to perform creative tasks.

Similarities and Boundaries Between Human Social Perception and Anthropomorphism of Robots

When evolving in social environments, one of the means to shape our social interactions is to quickly attribute internal states, based on as few cues as possible. Within human-human interactions, stereotype content model (Cuddy et al., 2007) posits that nearly all human counterparts may be categorized along two orthogonal dimensions relating, respectively, to: (1) their warmth (i.e., how friendly are they, how much do they care?) and (2) their competence (i.e., to what extent are they able to achieve their goals?). To have people—and presumably other social objects including brands (Kervyn et al., 2012, 2022)—categorized in such a simplified way gives direct clues regarding behaviors to adopt or judgments based on the emotional valence that is directly associated with the two dimensions combined (see, for instance, Bonetto et al., 2020). While we easily admire people (and possibly brands) that are perceived as both warm and able, we might feel contempt toward people on the other end of the scales—cold and unable, such as minorities (categorized as such) or homeless people.

Interestingly enough, attributing internal states in social interactions is the default mode for human agents, and even nonhuman entities may be imbued with mental states or "mind perception" (Wiese et al., 2017). In contrast to what ethologists could pinpoint as interpretation standards more than 100 years ago ("*In no case may we interpret an action as the outcome of the exercise of a higher psychical faculty, if it can be interpreted as the outcome of the exercise of one which stands lower on the psychological scale,*" Morgan, 1894), we humans appreciate animated objects or nonhuman entities, including animals (Darwin, 1859), geometrical objects (Heider & Simmel, 1944), but also deities and other nonhuman agents (Epley et al., 2008).

Facing a ubiquitous uncertainty about what exactly constitutes the real abilities or states of animals, objects, deities, or nonhuman agents, the best available proxy seems to be the human experience and knowledge itself (Epley et al., 2007). As such, *anthropomorphism* can then be evoked at two complementary levels. First, "simple physical constraints mean that humans have direct and immediate access to the phenomenological experience of being a human but do not have such immediate access to the phenomenological experience of any nonhuman agent" (Epley et al., 2007, p. 868). As it follows, when they then attribute mental states, abilities, or emotions to other nonhuman agents, they do so fairly easily on the basis of what they experience or know on their own, be it on individual, phenomenological level, or in general, as regards the human species.

As such, processes at the core of anthropomorphism may best be understood as inductive inference, that is, an extension of a sample experience of one's own cognitive system on unknown, uncertain, eventually nonhuman "objects" (Epley et al., 2007). This view is in line with several findings that relate to the heuristic nature of anthropomorphic and intentional inference. First, when under time pressure—that is, lacking time to override intuitive responses—participants in the first experiment of Rosset (2008) judged the prototypically accidental sentences (such as "She broke the vase" or "He bumped into a classmate in the hall") as done much more on purpose (22% of responses) than when they were unspeeded (only 15% of responses indicating actions done on purpose). The results thus supported the hypothesis of an *intentional bias*, where the default option was to consider that even accidental events might easily be interpreted as being intended by their actors.

In the same vein, Spatola and collaborators (Spatola et al., 2021a) investigated the degree of *intentional stance* (i.e., the propension to consider mental states as underlying explanations of an observed behavior) in two kinds of situations: isolated, where the iCub robot was about to do an action alone, or social, where the iCub robot was eventually in interaction with a human participant. Not surprisingly, the participants were much more likely to adopt an intentional stance (i.e., to ascribe mental states to iCub) in social situations (in which the robot interacted with a human partner) than in isolated ones. For example, the participants were more likely to admit that iCub "was trying to cheat by looking at the opponent's cards" (M = 63% out of 100) instead of being "unbalanced for a moment" in an interaction. In contrast, they were less prone to describe iCub as "fascinated by tool use" (M = 32% out of 100) instead of iCub "[grasping] the closest object" in an isolated action. These findings were in keeping with the "complexity approach" whereby mentalizing the action of an agent appears as a strategy to reduce complexity and the related uncertainty (Dennett, 1971). These results further underscore the inferential nature of psychological processes at the core of anthropomorphism (Epley et al., 2007).

Little doubt remains regarding the inferential nature of anthropomorphism processes and their similarities with human-human social perception. A thorough debate has been ongoing, however, as regards the number of dimensions pertaining to social robots' perception. A seminal work by Gray et al. (2007) has shown that mind perception—the fact of ascribing mental states to entities ranging here from animals through robots and living beings up to God—came along two scales termed Experience (the capacities to experience sensations like hunger or fear, but also personality, consciousness) and Agency—one's ability to self-control, plan, and communicate. Robots, represented here by Kismet, scored on the Agency dimension whereas being low on the Experience one. This result has been replicated in study by Gray and Wegner (2012) where computers *with* experience and, in parallel, humans *without* experience were considered as being the most unnerving, the least common within the known implicit schemes.

Some support for this pattern comes from neuroimaging studies where activations of brain areas were compared between situations where human participants observed another human versus a robot action (see Wiese et al., 2017, for a review). Two types of findings were of interest: First, as regards Agency, the brain areas engaged in understanding and planning actions were underactivated when observing robots as compared to observing humans, but this gap could be attenuated when participants were instructed to do so or after being allowed some familiarization time with robots. Second, as regards Experience and the ability to empathize or mentalize (i.e., to imagine how others feel or what they intend to do), the associated brain areas were activated only to a lesser degree when observing robots as compared to observing humans. This might explain why we show reluctant to attribute emotional states to robots, whereas we do not hesitate to ascertain their agency-related capacities.

Elaborating on previous dimensions and across four subsequent studies, Malle (2019) developed a three-way model of mind perception, integrating Affect (positive or negative, relating to emotional experience), Moral and social cognition (being under conscious control and thus relating to agency, although including one's and others' behavior), and Reality interaction, which includes transition from perception and cognition to action, including communication (and thus another part of agency). Spatola and collaborators (Spatola et al., 2021b) overviewed and completed the considered dimensions up to four, including sociability (close in meaning to Fiske's warmth and being determinant for robots' acceptance), intentionality (close to agency mentioned above and including rationality and intentionality of the robot), animacy (suggesting the human characteristics for nonhuman agents, such as real and alive), and disturbance, including items like scary, creepy, and uncanny.

Indeed, whereas ascribing mind makes an entities' actions meaningful (Waytz et al., 2010; Wiese et al., 2017) and may thus prove crucial for research in human-robot creativity, attributing internal states to nonhuman agents (anthropomorphizing them) obviously does not come without boundary conditions. Two cases are of interest: when the anthropomorphic model is not the best predictor for agent behavior (Epley et al., 2007; this second point is treated more in detail below), and when it is hard to categorize an agent as completely human or completely nonhuman (Cheetham et al., 2011, 2014). Previously, a linear relationship was supposed between the degree of anthropomorphic traits of a nonhuman agent and its positive evaluation. In contrast, the ***uncanny valley*** phenomenon describes the eeriness of robots which, while humanlike, produce at some stage more discomfort than a real social connection (Kätsyri et al., 2015; MacDorman et al., 2009; Mori, 1970). Based mostly on morphing paradigms—where images of fully human agents are progressively mixed to obtain fully nonhuman agents—some studies have suggested that, first, significant changes in mind ratings occur in a nonlinear manner, implying qualitative rather than quantitative progression (Cheetham et al., 2011). Second, it has been suggested that whereas categorization conflicts plateau at about 70% of humanness—which is coherent with Mori's prediction of uncanny valley appearance—this is exactly where participants experience the most cognitive conflicts to categorize agents as human versus nonhuman which, in turn, might result in feelings of eeriness of the targeted agent (Weis & Wiese, 2017). Another type of categorization conflict appears when the to-be-categorized agent presents both human (e.g., face) and nonhuman (e.g., body kinematics) traits. It can thus be recommended that ***for human-robot interaction, designs should be avoided that are prone to categorization conflict*** (uncertainty about being machine or human, Weis & Wiese, 2017) ***or to perceptual mismatch*** (partly human, partly nonhuman, Wiese et al., 2017). In contrast, it is of interest to ***design robots that can trigger flexible mental models***: anthropomorphic when novel and uncertain (or when human individuals are seeking for social connection and affiliation, Epley et al., 2007), and mechanistic when the nonhuman agent representation is already available and an effective task performance is the only aim (Wiese et al., 2017).

Cognitive and Affective Effects of Creative Engagement with Robots on Human Creativity

Beyond the production of anthropomorphic inferences, interacting creatively with a robot affects humans psychologically in a variety of ways. A small body of literature has been developing on this subject since the early 2010s, which we touched on descriptively in the previous chapter (by Chompunuch et al.; for reviews, see also Bossema et al., 2023; Lubart et al., 2021).

For example, we have seen in the previous chapter that the robot Shimon has been developed to play marimba and improvise jazz music with a human partner playing the piano (Hoffman & Weinberg, 2011). Interestingly, the results of this study showed that the pianists made visual contact with the robot only when it was changing rhythm, suggesting that this visual contact mainly aimed at timing the playing. However, the pianist's eye contact with the robot had an important influence on the audience. The latter judged the robot as playing better, as being more human, more inspired by his partner, and this better appreciation also extended to the pianist, who was also judged as being more inspired by his partner during eye contact. In other words, the same interaction was experienced and interpreted very differently by humans depending on their position. The players that directly interacted with the robot probably saw it as a tool, when the observers of the scene saw a coherent collaboration implying a communication.

The way in which interaction with the robot is experienced probably plays a central role regarding the positive effects of the latter on human creativity. A good illustration of this point comes from the study by Kahn et al. (2016). In this study, participants were asked to create a Zen Garden, with suggestions and questions displayed through a social robot or a PowerPoint presentation. Participants were twice more creative (i.e., used tools in an original way or created new types of effects in the garden) when interacting with a robot than with a computer presentation. But this increase in creativity mainly came from the fact that participants persevered longer in the presence of the robot, and thus had the opportunity to produce more creative acts. Several non-exclusive possible explanations exist to understand this effect. First, it is possible that the participants enjoyed interacting more with a robot than with a PowerPoint. Second, the robot might have produced a social facilitation effect akin to that of a human (Riether et al., 2012; Spatola et al., 2019a), namely an increase in performance due to the mere presence of another person (Zajonc, 1965; Guerin, 1993; Belletier et al., 2019). Third, the interaction with the robot might have been richer, more structured, or more similar with a usual interaction with a conspecific than the interaction with the PowerPoint. In any case, the interaction with the robot and its effect on the psychology of its human partner seems to be at the core of the effect.

This is probably why situations aiming at improving creativity through a collaboration between a human and a robot have mainly leveraged *social robots* defined as follows by Bartneck and Forlizzi (2004): "A social robot is an autonomous or semi-autonomous robot that interacts and communicates with humans by following the behavioral norms expected by the people with whom the robot is intended to interact." That is to say, social robots are (often implicitly) supposed to increase creativity because their usage is intuitive and because it imitates a natural interaction between humans. Consistently, when this goal is not achieved, the usage of a collaborative setting between a human and a robot does not improve creativity. For example, Alves-Oliveira et al. (2019) asked young adults to perform an open-ended drawing task either with the robot Nao or with a tablet. The limited capacity of the robot led participants to feel frustration (e.g., feeling that the robot ruined the drawing), and the robot's anthropomorphic appearance led them to build up expectations in terms of abilities that were in fact far greater than Nao's actual abilities. Hence, the creativity of the drawings did not increase when the robot was used compared to the tablet.

Different kinds of parameters in the robot's behavior can be used to improve the quality of the interaction and make it more humanlike. Fucinato et al. (2023) manipulated the type of vocal communication of the EZ robot. In this study, young adults were enrolled in an online creativity workshop and were divided into groups of three to four participants. Videos of the EZ robot to give instructions and guide the activities were then displayed to each group. The creativity task was a divergent thinking task in which participants had, for example, to invent a new chocolate product using images as inspiration sources. The EZ robot gave instructions and guidance either with a charismatic voice or with a non-charismatic voice through changes in prosody and pitch (the charismatic voice was 15 hertz higher and the range is 3 semitones larger than a reference voice, whereas the non-charismatic voice was lowered by 15 hertz and the range by 3 semitones). The robot with a charismatic voice was judged as more enthusiastic, charming, engaging, and passionate as well as more social. It also led to a better feeling of group dynamic and perceived creative performance. In other words, the robot with a charismatic voice improved the way the collaboration between humans was experienced. Moreover, the objective measures of creativity were also improved in terms of originality and elaboration. The gestures of the robots seem also to be of importance. Sandoval et al. (2022) asked adult participants to judge videos of a robot telling a creative story with or without moving. The moving robot was judged as more likable, safer, and having a higher animacy.

Altogether, the results we just reviewed are in line with the importance of the quality of the interaction with a social robot in order to improve creativity, and in particular the fact that this interaction should be natural, pleasant for the participant, and meet his or her expectations. Many of the studies that have been carried out on this subject have therefore used children as participants, probably because such interactions are easier to set up with this audience, especially if they are based on play.

Another reason for the predominance of studies involving children is the fact that children's creativity—i.e., their ability to generate novel, surprising, and valuable ideas—plays a crucial role in their learning, personal development, and well-being. For example, it enhances problem-solving skills, adaptability, self-expression, and even health (e.g., Alves-Oliveira et al., 2022; Gajda et al., 2017; Gardner & Gardner, 2008). However, creativity tends to decline between the ages of 8 and 10, influenced by various factors (Smith & Carlsson, 1985). For instance, school transitions can be stressful as children adjust to a new educational system. Additionally, cognitive development may impact creativity, as logical thinking becomes fully functional around the ages of 9 to 10 (Kim, 2011). The collaboration with a robot is therefore often suggested as a way of countering this decline in creativity by encouraging it in a stimulating environment with the hope of a spillover in other contexts.

Ali et al. (2020) tested whether children, aged 6 to 10, collaborating on a drawing with the Jibo robot were more creative than children collaborating with a tablet, even though both the robot and the tablet were using the same artificial intelligence (AI) to complete the drawing. The children reported having more fun with the robot, but found it less creative than the tablet, probably reflecting the fact that, as it has been observed with adults, *a* high level of expectation created by the robot's appearance but not achieved in terms of actual performance proves counterproductive. This interpretation seems consistent with the fact that, in this experiment, the children were no more creative when collaborating with the robot than with the tablet. In a subsequent study, Ali et al. (2021) directly manipulated the capacities of interaction displayed by the robot in three different tasks, measuring respectively verbal, figural, and constructional creativity. For example, figural creativity was measured using the same collaborative drawing as in the 2020 study, but the parameters of the IA were configured to produce drawings with high or low creativity. In all three tasks, children aged 5_10 were more creative when collaborating with the robot displaying a high level of creativity compared to the robot displaying a low level of creativity. It is noteworthy that the third task employed a different type of interaction in which the robot did not collaborate with the children but asked them reflexive questions.

Hence, the children were able to enhance their creativity through different types of interaction, but only when the robot was highly creative itself. This robotic creativity scaffolding was further tested in another task: a video game in which children aged 5–11 were asked to assemble a series of contraptions to move a character to a goal (Ali et al., 2022). The robot's dialogues differed between the creative and non-creative conditions. In the non-creative condition, its responses were loosely related to the context but did not aid gameplay. For instance, when a player used a trampoline, the robot might simply express its enjoyment of bouncing. In contrast, the creative condition featured dialogues designed to foster human creativity, based on Kahn et al.'s (2016) interaction patterns. These included demonstration (*Pushing the Limits*), inquiry (*Consider the Alternative*), and encouragement (*Validate Decision*). The robot was either physically present or virtually present in the game. The results showed that when the robot provided creativity support, its physical presence positively influenced children's creativity, particularly over time. However, whereas participants in the high-creativity conditions perceived both the virtual and physical robots as significantly more creative, they did not consider them to be more helpful. Altogether, the experiments developed by this research team emphasize the importance of the actual capabilities of the robots used with children, probably because if these capabilities don't reach the level children anticipate, they don't take advantage of the robot as a model or guide for improving their creativity.

Elgarf et al. (2021) also compared a creative robot and a non-creative robot in the setting of a storytelling collaboration with children aged from 5 to 10. The Furhat robot was used and its creativity was manipulated through the number, the originality, and the expressivity of the ideas to continue the story. However, the robot's level of creativity did not influence the children's creativity. The authors noted that some children taking part in the experiment were frustrated by the robot's interruptions and rejected its suggestions, sometimes expressing annoyance. Others passively accepted the robot's ideas, reducing their own contributions to simple acknowledgments. In short, the problem was that the robot's behavior was not adaptive to the children's one.

To address this issue, Elgarf et al. (2022) adapted the Regulatory Focus Theory (Higgins, 1998) to the context of child–robot interaction. According to this theory, individuals can pursue goals during task performance either with a promotion focus, which motivates goal achievement through positive anticipation, or with a prevention focus, which motivates behavior by avoiding negative outcomes. Research suggests that promotion focus, linked to excitement and happiness, enhances creativity more than prevention focus, which is associated with fear and relief. In consequence, Elgarf et al. (2022) asked children to perform a collaborative storytelling task with a robot, with a goal to attain while the robot made positive comments (e.g., "*I am so excited to do this!*") or with a goal to avoid while the robot made negative comments (e.g., "*I am so scared of the explosion!*"). Results showed that a promotion-focused interaction enhanced fluency and average originality, but the robot's interventions often temporarily reduced originality, likely due to the children's negative perception of the interruptions. Nonetheless, the use of this model opens an avenue to better consider the way emotions might play a role in the creative collaboration between children and robots.

In a third study, Elgarf et al. (2024) used a robot with a Large Language Model (ChatGPT 3) trained to be creative or non-creative, and which behaved in a collaborative or a non-collaborative way. The robot's creative behavior helped children generate more ideas with greater detail and variety, but did not increase originality. Children preferred the non-collaborative conditions because they felt more freedom and autonomy in their storytelling. In contrast, the robot's interference in collaborative conditions led to frustration, as children desired more control over their stories. To sum up, this line of research suggests that robot creativity alone does not enhance children's creativity in storytelling; instead, children may prefer more autonomy and less interference from the robot. Emotional and motivational factors, such as those explored through Regulatory Focus Theory, could play a significant role in shaping children's responses to the robot's behavior, with promotion-focused interactions enhancing fluency and originality, though still subject to frustration from interruptions. Ultimately, the studies highlight *the importance of adapting robots' behaviors to children's preferences and emotional states to foster more effective and enjoyable collaborative creativity.*

Calvo-Barajas and Castellano (2022) made pioneering work concerning the long-term effects of interacting with a robot. They asked children aged from 9 to 12 years to play an interactive storytelling game with the Furhat robot in two sessions separated by one week. The study incorporated both verbal and non-verbal robot behaviors to guide and engage children during interactions. Non-verbal behaviors included head movements like nodding and shaking, as well as facial expressions to convey emotions such as happiness and surprise. Verbal behaviors were aimed at encouraging children to tell stories, provide feedback, generate ideas, and engage in general conversation. The study was run online via a Zoom video call. Repeated interactions with a virtual robot improved children's creativity in Fluency, Flexibility, and Elaboration, but did not affect Originality. Children told longer and more detailed stories in the second session, likely due to greater familiarity with the task and the robot. Originality remained unchanged, possibly because children used the same characters and objects in both sessions, leading to a lack of novelty in their stories. Children also became more likely to incorporate the robot's ideas into their stories and exhibited more verbal social behaviors, indicating a growing rapport with the robot over time. In sum, repeated interactions with robots seem to have an influence on children's creativity, particularly in fluency, flexibility, and elaboration. However, the lack of improvement in originality and the limited novelty across sessions raise questions about the long-term effectiveness of robot interventions in fostering truly innovative thinking, and more research is needed on this topic.

Alves-Oliveira et al. (2021) took a different approach to the different studies we have reviewed so far. Instead of using a premade robot, they developed a robot called YOLO designed to improve creative behaviors during unstructured play and storytelling. This development was made by involving children in order to develop a type of interaction that would be natural and intuitive, with no turn-taking restrictions. Although YOLO was developed to be a social robot, the designers did not choose to give it an anthropomorphic appearance, to set realistic expectations for the robot's capabilities. The creativity module of the robot allows it to display similar movements as the ones initiated by children at the beginning and the end of the play session (to emulate convergent thinking) and to exhibit different movements from the ones created by children in the middle of the play session (to emulate divergent thinking). The robot communicates basically using lights with different colors. Altogether, the characteristics of this robot correspond to those we have identified so far in this chapter as enabling an increase in children's creativity through positively experienced interaction, namely an intuitive and non-disappointing use. Consistently, children aged between 7 and 9 who played with this robot showed greater creativity when the robot used its own creativity module (Alves-Oliveira et al., 2020).

In summary, the reviewed studies reveal that the quality and nature of interactions with social robots significantly influence human creativity, especially in children. The effectiveness of these interactions depends on a range of factors, including ***the robot's actual creative capabilities in relation to the expectations the user has formed based on its appearance, the adaptability of its behavior, the emotional and motivational context in which it engages with users***, but also probably the level of perceived agentivity attributed to it (Moruzzi, 2022). While some experiments demonstrated enhanced performance in creative tasks, others highlighted that unmet expectations or overly intrusive behavior can lead to frustration and diminished originality. Moreover, the impact of social cues such as vocal charisma, non-verbal gestures, and even the physical presence of the robot underscores the complex role of human-robot interaction in creative collaboration. Despite promising findings, the mixed results and limitations noted in several studies suggest that further research is needed to fully understand and optimize the potential of robots as creative partners. Finally, social robots are generally proposed for this type of interaction, as they enable a more intuitive interaction, closer to that between humans. This raises the question of whether or not the classic social effects usually observed in creative human-human interactions can be found in hybrid human/robot teams.

General Discussion and Conclusion

We have seen in this chapter that human beings can cognitively treat interactions with a robot as "social" interactions. However, this transition from a simple technical interaction to a sociotechnical interaction depends heavily on the conditions being favorable, i.e., that the robot's appearance and behavior maximize the perception of agentivity, sociability, and autonomy in its human partners, while reducing the feeling of distrust that could be provoked by an appearance or behavior too close to that of a human. Furthermore, the few studies that have focused more specifically on collaborative human-robot interaction demonstrate the crucial importance of human expectations of the robot, both in terms of technical and social skills (ability not to interrupt, to respond appropriately, etc.).

Taken together, these studies raise the question of the extent to which the models we have shown all along this book to address the complexity of collective creative activities between humans can be transferred or adapted to human-machine (computer or system) environments, such as human-robot interactions. For example, the stereotype content model (Cuddy et al., 2007), based on the two dimensions of warmth and competence that we discussed earlier, and which describes how we form impressions of the individuals we encounter, has been applied to impression formation regarding robots (e.g., Mieczkowski et al., 2019; Spatola et al., 2019b). Thus, humans tend to interpret interactions with machines, particularly robots, in a social way, when certain conditions of social presence are met (e.g., perceived agency, apparent autonomy, non-mechanical behavior). However, are existing models of human collaboration and/or human creativity sufficient to account for the dynamics of human-machine, human-AI and/or human-robot interactions during creative activities?

All along this book, we have explored the many faces of collaboration in creative contexts, first analyzing collective creative activities in situations based on human interactions, then looking at interactions between humans and artificial systems. In the last chapters, we saw the emergence of a complex ecology in which the roles between human agents and artificial agents are evolving (see, for instance, Lubart, 2005; Lubart et al., 2021), and even the boundaries between humans, tools, and partners can become increasingly porous. In particular, three main modes of collaboration between humans and artificial intelligence have been described in the previous chapter (by Chompunuch et al.): AI as an *aid*, AI as a *partner*, and AI as a *primary creator*. Each of these modes redefines roles, responsibilities, and drivers of creativity, shifting the balance of control from humans to a form of computational autonomy. These dynamics also suggest that AI could be both a catalyst and a disruptor in creative activities. As a *catalyst*, AI can amplify human imagination, generate novel associations, and accelerate ideation by offering unexpected or alternative perspectives. It can enable individuals and groups to explore vast combinatorial spaces quickly, pushing the boundaries of traditional creative processes. However, as a *disruptor*, AI challenges established notions of authorship, originality, and creative control. It may destabilize creative identities and professional practices, as it has been pointed out by Vinchon et al. (2023). To navigate this dual potential effectively, in line with an ergonomics approach (Bonnardel, 2006), it is essential to thoroughly analyze the needs, capabilities, and limitations of each human agent (or user), taking into account their evolution, for instance, regarding the acquisition of expertise in the creative domain. We argue that collaboration with AI must be adaptive, responding not only to the complexity of the creative task, but also taking into account the specificities of each human agent and the situational context. An interesting perspective is provided by the Hybrid Intelligence Technology Acceptance model (HI-TAM), which offers an innovative framework for training and adopting generative design assistants, facilitating human-AI co-creation, as analyzed in the field of architectural design (Mao et al., 2024). It aims at enabling continuous customization and improving the psychological acceptance of AI. Complementary research will especially require in-depth *reflection on the kind of collaboration needed, when it should be provided, and how it should be provided.*

This also brings us back to a question evoked at the beginning of this chapter: can we really talk of *collaboration* (in the strong sense of the term) when one of the "partners" is an artificial entity, often limited in its cognitive autonomy, contextual interpretation, and creative initiative? This question raises fundamental issues: does collaboration necessarily imply shared intentionality as well as a mutual understanding of goals, constraints, and emotional and social dynamics? Instead of collaboration, would it be more relevant to speak of *cooperation*, which refers to a mode of joint activity in which individuals work toward a shared objective by dividing tasks and/or responsibilities? Or must we accept a redefinition, perhaps weakened, of the very concept of collaboration when it comes to interactions with artificial intelligence or robots?

Moreover, is it or will it be realistic or even desirable to consider that human and artificial agents can fully be engaged in a genuine process of *co-creation between humans and artificial agents*? As defined by Vinchon et al. (2023), it occurs when humans and AI work together to create new forms of art, design, and innovation, with both contributing without one dominating the other. This hybridization of creative agents raises fundamental issues (e.g., technical, cognitive, social, and ethical), which were addressed and analyzed in this book from different perspectives, but which will also require further research.

Finally, creativity may no longer be viewed as an exclusively human or technological domain. It can be considered as a "territory of interaction," where relational qualities, cognitive, emotional, and social models as well as sociotechnical architectures converge to produce new processes and/or forms of creation. By accepting this complexity and keeping alive the tension between collaborative illusion and effective collaboration, we are paving the way for a profound redefinition of what it means to create together—humans and machines alike—acknowledging that creativity can be based on new relationships, where the dialogue between biological and artificial intelligence transforms the way we conceive, imagine, and collaborate.

References

Ali, S., Devasia, N. E., & Breazeal, C. (2022, June). Escape! Bot: Social robots as creative problem-solving partners. In *Proceedings of the 14th conference on creativity and cognition* (pp. 275–283).

Ali, S., Park, H. W., & Breazeal, C. (2020, November). Can children emulate a robotic non-player character's figural creativity? In *Proceedings of the annual symposium on computer-human interaction in play* (pp. 499–509).

Ali, S., Park, H. W., & Breazeal, C. (2021). A social robot's influence on children's figural creativity during gameplay. *International Journal of Child-Computer Interaction, 28*, 100234.

Alves-Oliveira, P., Arriaga, P., Paiva, A., & Hoffman, G. (2021). Children as robot designers. In *Proceedings of the 2021 ACM/IEEE international conference on human-robot interaction* (pp. 399–408).

Alves-Oliveira, P., Arriaga, P., Xavier, C., Hoffman, G., & Paiva, A. (2022). Creativity landscapes: Systematic review spanning 70 years of creativity interventions for children. *Journal of Creative Behavior, 56*(1), 16–40.

Alves-oliveira, P., Tulli, S., Wilken, P., Merhej, R., Gandum, J., & Paiva, A. (2019). *Sparking creativity with robots: A design perspective*. https://doi.org/10.31219/osf.io/za5h8

Alves-Oliveira, P., Arriaga, P., Cronin, M. A., & Paiva, A. (2020, March). Creativity encounters between children and robots. In Proceedings of the 2020 ACM/IEEE international conference on human-robot interaction (pp. 379-388).

Bartneck, C., & Forlizzi, J. (2004). A design-centred framework for social human-robot interaction. In *RO-MAN 2004. 13th IEEE international workshop on robot and human interactive communication (IEEE catalog no. 04TH8759)* (pp. 591–594). IEEE.

Belletier, C., Normand, A., & Huguet, P. (2019). Social-facilitation-and-impairment effects: From motivation to cognition and the social brain. *Current Directions in Psychological Science, 28*(3), 260–265.

Bonetto, E., Pichot, N., Girandola, F., & Bonnardel, N. (2020). The normative features of creativity: Creative individuals are judged to be warmer and more competent. *Journal of Creative Behavior, 55*, 649–660. https://doi.org/10.1002/jocb.477

Bonnardel, N. (2006). Entrée: Ergonomie: Communication Homme-Machine. In S. Mesure & P. Savidan (Eds.), *Dictionnaire des Sciences Humaines* (pp. 386–388). PUF.

Bonnardel, N. (2024). Design activities. In F. Darbellay (Ed.), *Elgar encyclopedia of interdisciplinarity and transdisciplinarity*. Edward Elgar Publishing.

Bonnardel, N., & Didier, J. (2016). Enhancing creativity in the educational design context: An exploration of the effects of design project-oriented methods on students' evocation processes and creative output. *Journal of Cognitive Education and Psychology, 15*(*1*), 80–101. https://doi.org/10.1891/1945-8959.15.1.80

Bonnardel, N., & Didier, J. (2020). Brainstorming variants to favor creative design. *Applied Ergonomics, 83*, 102987. https://doi.org/10.1016/j.apergo.2019.102987

Bonnardel, N., Girandola, F., Bonetto, E., & Lubart, T. (Eds.). (2023). *La Créativité en Situations: Théories et Applications*. Dunod.

Bonnardel, N., & Pichot, N. (2020). Enhancing collaborative creativity with virtual dynamic personas. *Applied Ergonomics, 82*, 102949. https://doi.org/10.1016/j.apergo.2019.102949

Bossema, M., Allouch, S. B., Plaat, A., & Saunders, R. (2023, August). Human-robot co-creativity: A scoping review: Informing a research agenda for human-robot co-creativity with older adults. In *2023 32nd IEEE international conference on robot and human interactive communication (RO-MAN)* (pp. 988–995). IEEE.

Brangier, E., Bornet, C., Bastien, J. M. C., Michel, G., & Vivian, R. (2011). Mesure de la capacité des personas à générer des idées dans la conception de projets WEB. *Le Travail Humain, 75*(2), 121–145.

Calvo-Barajas, N., & Castellano, G. (2022). "I have an idea!" Enhancing children's verbal creativity through repeated interactions with a virtual robot. In *Proceedings of the 22nd ACM international conference on intelligent virtual agents* (pp. 1–8).

Cheetham, M., Suter, P., & Jäncke, L. (2011). The human likeness dimension of the "uncanny valley hypothesis": Behavioral and functional MRI findings. *Frontiers in Human Neuroscience, 5*, 126.

Cheetham, M., Suter, P., & Jäncke, L. (2014). Perceptual discrimination difficulty and familiarity in the uncanny valley: More like a "Happy Valley". *Frontiers in Psychology, 5*, 1219.

Corazza, G. E. (2025). *The cyber-creativity process. How humans co-create with artificial intelligence*. Palgrave Macmillan.

Cuddy, A. J., Fiske, S. T., & Glick, P. (2007). The BIAS map: Behaviors from intergroup affect and stereotypes. *Journal of Personality and Social Psychology, 92*(4), 631–648.

Darwin, C. (1859/2013). *On the origin of species*. Kent/In libro veritas.

Dennett, D. C. (1971/2010). *Content and consciousness*. Routledge.

Détienne, F., Baker, M., & Burkhardt, J.-M. (Eds.). (2012a). Perspectives on quality of collaboration in design. *CoDesign, 8*(*4*), 197–199.

Détienne, F., Baker, M., & Burkhardt, J.-M. (2012b). Quality of collaboration in design meetings: Methodological reflexions. *CoDesign: International Journal of CoCreation in Design, 8*(4), 247–261.

Elgarf, M., Calvo-Barajas, N., Alves-Oliveira, P., Perugia, G., Castellano, G., Peters, C., & Paiva, A. (2022). "And then what happens?" Promoting children's verbal creativity using a robot. In *2022 17th ACM/IEEE international conference on human-robot interaction (HRI)* (pp. 71–79). IEEE.

Elgarf, M., Salam, H., & Peters, C. (2024). Fostering children's creativity through LLM-driven storytelling with a social robot. *Frontiers in Robotics and AI, 11*, 1457429.

Elgarf, M., Skantze, G., & Peters, C. (2021). Once upon a story: Can a creative storyteller robot stimulate creativity in children? In *Proceedings of the 21st ACM international conference on intelligent virtual agents* (pp. 60–67).

Epley, N., Akalis, S., Waytz, A., & Cacioppo, J. T. (2008). Creating social connection through inferential reproduction: Loneliness and perceived agency in gadgets, gods, and greyhounds. *Psychological Science, 19*(2), 114–120.

Epley, N., Waytz, A., & Cacioppo, J. T. (2007). On seeing human: A three-factor theory of anthropomorphism. *Psychological Review, 114*(4), 864–886.

Fucinato, K., Niebuhr, O., Nørskov, S., & Fischer, K. (2023). Charismatic speech features in robot instructions enhance team creativity. *Frontiers in Communication, 8*, 1115360.

Gajda, A., Karwowski, M., & Beghetto, R. A. (2017). Creativity and academic achievement: A meta-analysis. *Journal of Educational Psychology, 109*(2), 269.

Gardner, H., & Gardner, E. (2008). *Art, mind, and brain: A cognitive approach to creativity*. Basic Books.

Gray, H. M., Gray, K., & Wegner, D. M. (2007). Dimensions of mind perception. *Science, 315*(5812), 619–619.

Gray, K., & Wegner, D. M. (2012). Feeling robots and human zombies: Mind perception and the uncanny valley. *Cognition, 125*(1), 125–130.

Guerin, B. (1993). *Social facilitation*. University Press.

Heider, F., & Simmel, M. (1944). An experimental study of apparent behavior. *The American Journal of Psychology, 57*, 243–259.

Higgins, E. T. (1998). Promotion and prevention: Regulatory focus as a motivational principle. In *Advances in experimental social psychology* (Vol. 30, pp. 1–46). Academic Press.

Hoffman, G., & Weinberg, G. (2011). Interactive improvisation with a robotic marimba player. *Autonomous Robots, 31*(2), 133–153.

Kahn, P. H., Kanda, T., Ishiguro, H., Gill, B. T., Shen, S., Ruckert, J. H., & Gary, H. E. (2016, March). Human creativity can be facilitated through interacting with a social robot. In *2016 11th ACM/IEEE international conference on human-robot interaction (HRI)* (pp. 173–180). IEEE.

Kätsyri, J., Förger, K., Mäkäräinen, M., & Takala, T. (2015). A review of empirical evidence on different uncanny valley hypotheses: support for perceptual mismatch as one road to the valley of eeriness. *Frontiers in Psychology, 6*, 390.

Kervyn, N., Fiske, S. T., & Malone, C. (2012). Brands as intentional agents framework: How perceived intentions and ability can map brand perception. *Journal of Consumer Psychology, 22*(2), 166–176. https://doi.org/10.1016/j.jcps.2011.09.006

Kervyn, N., Fiske, S. T., & Malone, C. (2022). Social perception of brands: Warmth and competence define images of both brands and social groups. *Consumer Psychology Review, 5*(*1*), 51–68.

Kim, K. H. (2011). The creativity crisis: The decrease in creative thinking scores on the Torrance Tests of Creative Thinking. *Creativity Research Journal, 23*(4), 285–295.

Lubart, T. (2005). How can computers be partners in the creative process: Classification and commentary on the special issue. *International Journal of Human-Computer Studies, 63*(*4–5*), 365–369.

Lubart, T. (Ed.). (2018). *The creative process: Perspectives from multiple domains*. Palgrave Macmillan.

Lubart, T., Esposito, D., Gubenko, A., & Houssemand, C. (2021). Creativity in humans, robots, humbots. Creativity. *Theories–Research-Applications, 8*(*1*), 23–37.

MacDorman, K. F., Green, R. D., Ho, C. C., & Koch, C. T. (2009). Too real for comfort? Uncanny responses to computer generated faces. *Computers in Human Behavior, 25*(3), 695–710.

Malle, B. F. (2019). How many dimensions of mind perception really are there? In *Proceedings of the annual meeting of the cognitive science society* (Vol. 41).

Mao, Y., Rafner, J., Wang, Y., & Sherson, J. (2024). HI-TAM, a hybrid intelligence framework for training and adoption of generative design assistants. *Frontiers in Computer Science, 6*, 1460381.

Mieczkowski, H., Liu, S. X., Hancock, J., & Reeves, B. (2019, March). Helping not hurting: Applying the stereotype content model and BIAS map to social robotics. In *2019 14th ACM/IEEE international conference on human-robot interaction (HRI)* (pp. 222–229). IEEE.

Morgan, C. (1894). The homing of limpets. *Nature, 51, 127.*

Mori, M. (1970). The uncanny valley. *Energy, 7*, 33–35.

Moruzzi, C. (2022). Creative agents: Rethinking agency and creativity in human and artificial systems. *Journal of Aesthetics and Phenomenology, 9*(2), 245–268.

Riether, N., Hegel, F., Wrede, B., & Horstmann, G. (2012, March). Social facilitation with social robots?. In *Proceedings of the seventh annual ACM/IEEE international conference on Human-Robot Interaction* (pp. 41–48).

Robert, J. M., & Brangier, E. (2012). Prospective ergonomics: Origin, goal, and prospects. *Work, A Journal of Prevention, Assessment & Rehabilitation, 41*, 5235–5242.

Rosset, E. (2008). It's no accident: Our bias for intentional explanations. *Cognition, 108*(3), 771–780.

Sandoval, E. B., Sosa, R., Cappuccio, M., & Bednarz, T. (2022). Human–robot creative interactions: Exploring creativity in artificial agents using a storytelling game. *Frontiers in Robotics and AI, 9*, 695162.

Smith, G., & Carlsson, I. (1985). Creativity in middle and late school years. *International Journal of Behavioral Development, 8*(3), 329–343.

Spatola, N., Anier, N., Redersdorff, S., Ferrand, L., Belletier, C., Normand, A., & Huguet, P. (2019a). National stereotypes and robots' perception: The "made in" effect. *Frontiers in Robotics and AI, 6*, 21.

Spatola, N., Belletier, C., Chausse, P., Augustinova, M., Normand, A., Barra, V., et al. (2019b). Improved cognitive control in presence of anthropomorphized robots. *International Journal of Social Robotics, 11*(3), 463–476.

Spatola, N., Kühnlenz, B., & Cheng, G. (2021a). Perception and evaluation in human–robot interaction: The Human–Robot Interaction Evaluation Scale (HRIES)—A multicomponent approach of anthropomorphism. *International Journal of Social Robotics, 13*(*7*), 1517–1539.

Spatola, N., Marchesi, S., & Wykowska, A. (2021b). The intentional stance test-2: How to measure the tendency to adopt intentional stance towards robots. *Frontiers in Robotics and AI, 8*, 666586.

Vincent-Lancrin, S., Gonzalez-Sancho, C., Bouckaert, M., de Luca, F., Fernández-Barrerra, M., Jacotin, G., Urgel, J., & Vidal, Q. (2019). *Fostering students' creativity and critical thinking: What it means in school.* OECD Publishing. https://doi.org/10.1787/62212c37-en

Vinchon, F., Lubart, T., Bartolotta, S., Gironnay, V., Botella, M., Bourgeois-Bougrine, S., Burkhardt, J.-M., Bonnardel, N., Corazza, G. E., Glăveanu, V., Hanchett Hanson, M., Ivcevic, Z., Karwowski, M., Kaufman, J. C., Okada, T., Reiter-Palmon, R., & Gaggioli, A. (2023). Artificial intelligence & creativity: A manifesto for collaboration. *Journal of Creative Behavior, 57*(*4*), 472–484. https://doi.org/10.1002/jocb.597

Waytz, A., Cacioppo, J., & Epley, N. (2010). Who sees human? The stability and importance of individual differences in anthropomorphism. *Perspectives on Psychological Science, 5*(3), 219–232.

Weis, P., & Wiese, E. (2017). Cognitive conflict as possible cause for the uncanny valley. In *Proceedings of human factors and ergonomics society.*

Wiese, E., Metta, G., & Wykowska, A. (2017). Robots as intentional agents: Using neuroscientific methods to make robots appear more social. *Frontiers in Psychology, 8*, 1663.

Woodman, R. W., Sawyer, J. E., & Griffin, R. W. (1993). Toward a theory of organizational creativity. *Academy of Management Review, 18*(2), 293–321.

Zajonc, R. B. (1965). Social Facilitation: A solution is suggested for an old unresolved social psychological problem. *Science, 149*(3681), 269–274.

Index

A

Activities, vii, viii, 1–33, 45–51, 53, 54, 58, 59, 63, 65–68, 76, 79, 91, 94, 96–98, 135, 136, 138, 155, 157–160, 162, 180, 183, 189, 190, 192, 196, 222, 225, 226, 251, 252, 268, 276–279, 285, 291, 292

Adversarial collaboration, 170

Affects, 30, 52, 56–58, 76, 95, 102, 167, 168, 190, 199, 282, 284, 289

Affordances, 5, 27, 32, 194, 205–208, 218

Agency, 9, 49, 95, 99, 248, 279, 282, 291

Algorithm, 191, 266

Annotation, 216, 218, 219, 224–230, 232, 233, 236, 237, 239–241, 278

Anthropomorphism, 280–283

Argumentation, 23, 27, 52, 55, 56, 58, 61, 63, 67, 277

Artifact-mediated interaction, 216

Artifacts, 5, 9, 24, 166–169, 192, 195, 200, 202, 206, 207, 213, 215–220, 240, 247, 249, 262

Artificial intelligence (AI), vii, viii, 6, 68, 108, 148, 200, 202, 208, 209, 245–270, 279, 286, 291–293
- applications, 268
- as Helper, 246, 248–252, 254–256, 269
- as Partner, 246, 250, 258, 259, 269, 291
- as Principal Creator, 245–247, 250, 262, 264, 266, 269

Authors, 4, 8, 81, 98, 131, 133, 135, 160, 163, 164, 172, 178, 191, 200–202, 204, 205, 208, 210, 255, 260, 278, 288

Avatars, 27–32, 84

B

Behavioral coding scheme, 138

Behaviour, 48, 59, 67

N. Bonnardel (ed.), *Collaborations in Creativity*, The Seven C's of Creativity, https://doi.org/10.1007/978-3-032-07753-0

Brain, 190–192, 194, 199, 205, 251, 278, 282
Brainstorming, 9, 19, 21, 32, 52, 79, 83–89, 101, 107, 108, 129, 160, 161, 206, 251

C
Characters, 190, 199, 203, 204, 206, 207, 225, 253, 260, 287, 289
Children's creativity, 286–290
Citizen-driven inquiry, 214
Coach, 136, 198, 199, 201, 202, 208, 251–252
Co-creativity, 6–10, 12, 24, 32, 247, 256
Codevelopment, 6, 20, 193, 198–205, 209
COFOR model, 12–14, 24, 81, 277
Cognition, 6, 128, 131, 189–209, 278, 282
Cognitive diversity, 10, 83, 86, 94, 107
Collaboration, vii, 1–33, 45–68, 75, 129, 155–184, 191, 195–205, 216, 245–270, 275–293
  appraisal of, 64–66
  in creativity, 20, 276
  definition, 47–49, 170, 192
  dimensions, 23, 63–65, 67
  quality assessment grid, 277
  quality of, 6, 23, 28, 29, 46, 54, 57, 59, 60, 62–67, 158, 164, 277
Collaborative creativity, vii, 27, 47, 56, 289
Collaborative inhibition/process loss, 161
Collages, 199, 200, 202, 207
Collective activity, 47–49, 53, 67
Collective creative activities, 4, 7, 9–10, 12–21, 32, 79, 276, 277, 291
Collective creativity, 4, 6, 10, 17, 32, 33, 77–84, 90, 94, 95, 107, 208, 213–242, 278
Collectively, 9, 14, 24, 87, 90, 129, 207, 230, 233, 268, 278
Computational neuroscience, 191
Computer supported collaborative learning (CSCL), 47, 63
Conflict, 4, 19, 55, 56, 58, 60, 61, 100–103, 107, 162, 167, 178–181, 190, 195, 283
  socio-cognitive, 58
Conformity, 11, 81–82
Connected, 23, 230, 277
Constraints, 2, 4, 6, 17, 23, 24, 31, 51, 68, 86, 87, 96–101, 105, 107, 179, 197, 229, 250, 263, 264, 280, 292
Constructional creativity, 287
Constructive, 20, 25, 32, 57–58, 66, 67, 101, 107, 170, 178
Contributions, vii, viii, 7, 8, 10, 16, 20, 26, 50, 51, 54–57, 63, 77, 104, 134, 180, 202, 207, 230, 246, 247, 250, 263, 269, 277, 288
Convergent thinking, 27, 83, 252, 254, 289
Cooperation, 6–9, 13, 14, 24, 32, 47–50, 57, 67, 170, 179, 189, 221, 292
Coordination, 7, 8, 13, 14, 27, 47–49, 51–52, 67, 102, 107, 259
Creative activities, vii, viii, 1–33, 76, 79, 91, 96–98, 222, 251, 252, 276–278, 291
Creative problem-solving, 101, 128–130, 134, 135, 148, 155
Creative process, vii, 5, 6, 8, 9, 11, 26, 53, 76, 77, 81, 82, 96, 100, 129–131, 195, 197, 200, 218, 222, 245–256, 258, 262, 264–267, 269, 291

Creativity, vii, 2, 46, 75–108, 127–149, 156, 191, 213–242, 245, 249–251, 276, 284–290
Cultural diversity, 86, 107
Cultural spaces, 216
Culture, 27, 46, 58–62, 68, 77, 79, 93, 166, 208, 215, 261, 277

D
Dark-Room, 192, 193
Data engagement, 219, 223, 224
De Bono's thinking hats, 22, 23
Debrief, 135, 136, 200
Design, vii, 2, 45–68, 89, 171, 209, 216, 220–223, 252, 277
  collaboration in, 46, 56, 63
Designed serendipity, 175–176, 178
Design-fiction, 21, 24–27, 32, 278
Dialectic knowledge creation, 155–184
Digital system, 7
Discourse, 7, 59, 164, 174, 190
Divergent thinking, 5, 11, 82, 129, 252, 285, 290
Drawings, 24, 54, 55, 200–202, 206, 207, 216, 223, 228, 248, 265, 285–287
Dynamic personas, 27–30, 33, 278

E
Embedded cognition, 193, 194, 206, 278, 279
Embodied cognition, vii, viii, 6, 189–209, 278
Enactive cognition, 193, 194, 278
Evaluation, 6, 11, 19, 21, 31, 53, 55, 59, 60, 62, 66, 67, 87, 88, 105, 108, 129–131, 134, 135, 142–144, 149, 158, 162, 164, 170, 178, 179, 209, 249, 255, 269, 278, 283
  appraisal, 60, 64, 66
Expectations, 17, 82, 84, 95, 96, 161, 192, 263, 285, 286, 289–291
Extended cognition, 207
Externalization, 201, 206, 241

F
Faces, 4, 12, 31, 51, 55, 163, 171, 181, 199, 200, 202, 206, 207, 283, 291
Figural creativity, 287
Filmmakers, 6, 198, 209
Five A's framework, 218
Flow, 11, 90, 100, 131, 206, 259
Foresight, 25, 26, 182, 209
4E perspective, 193, 205–208, 278

G
Game design, 263
GREENHOUSE, 193, 200–209
Group, vii, 4, 48, 75, 82–83, 135, 156, 189, 217, 228–230, 258, 277
Group activities, 48

H
Historical statistical tables, 213–242, 278
Human-AI collaboration, 246–250, 261, 267, 269
Human-AI interaction, 247, 256, 291
Human and AI partnership, 256, 258
Human-centered creativity, 246, 249–252, 269
Human collaboration, 277, 291
Human-human collaboration, 6, 20–27, 279
Human-machine collaboration, 256, 267, 291

Human-robot interaction (HRI), 248, 250, 279, 283, 290, 291
Human-social robot collaboration, 279
Human-system collaboration, 1–33, 279
Hybrid Intelligence Technology Acceptance model, 248, 292

I

Idea elaboration, 129, 131, 135, 142, 158, 161–162, 164
Idea evaluation, 129–131, 134, 135, 142–144, 149, 158, 255
Idea generation, 11, 30, 32, 52, 84–87, 92, 129–131, 134, 135, 141, 142, 144, 149, 156, 158–161, 171, 172, 175, 209, 249, 254, 255
Idea selection, 11, 129–131, 144–145, 156, 249
Immersion, 27, 204
Implications of AI, viii
Improvisation, 202, 259
Intentional stance, 281
Interdependency, 50
Intuitive, 202, 207, 208, 281, 285, 289, 290

L

Large language model, 148, 288
Leadership, 19, 79, 83, 90–94, 102–105, 107, 108

M

Mediation, 164, 242
  graphic representations, 53, 54
Mental representations, 3, 13, 14, 191, 192, 194, 202
Methods, 7, 12, 20–29, 32, 33, 51, 52, 60, 62–65, 67, 68, 77, 84, 100, 101, 135, 157, 158, 169, 170, 172, 175, 176, 195, 198, 208, 223, 230, 233, 240, 278
Mind perception, 280, 282
Mini-c creativity, 218, 278
Models, 5, 6, 12–21, 24, 26, 32, 33, 76–79, 81, 85, 86, 89, 96–100, 102, 128–133, 144, 148, 156, 195, 197, 209, 218, 219, 246, 256, 259, 262, 263, 277, 280, 282, 283, 287, 288, 291, 293
Modes of communication, 25
Multi-level interaction/collaboration, 6, 155–184
Museum-inspired design, 220–223

N

Narrative, 24, 25, 67, 190, 193, 198, 200, 201, 205, 207, 208, 214, 215, 246, 261, 263, 278

O

Open science and interdisciplinary research, 156, 278
Orality, 199, 200, 208
Organizational creativity, 6, 75–108

P

Participatory knowledge creation, 278
Peer review, 156–158, 160, 162–165, 180
Personas, 23, 27–30, 65, 278
Perspectives, vii, 4, 6, 10, 12, 15, 16, 20–22, 24–26, 32, 46, 53, 65–68, 76–78, 80–83, 91, 98, 99, 101, 104, 105, 131–135, 144, 156–158, 163, 165, 170–172,

175–182, 184, 191, 193–195, 202, 203, 205–209, 216, 218–221, 225, 239, 240, 258, 265, 278, 279, 291, 292
Plot, 190
Problem construction, 129–131, 134, 135, 141–144, 149
Productive, 56–58, 66, 67, 102
Proteus Effect, 27, 30, 31
Psychology, 3, 23, 25, 29, 75, 80, 105, 156, 157, 191, 193, 198, 285

R
Real-life, 6
Real-life collective creative situations, 278
Real-world contexts, 9
Reflexive, 57–58, 67, 68, 192, 287
Regulatory Focus Theory, 288
Responsible Research Innovation (RRI), 182
Robot, viii, 6, 81, 245–270, 275–293
Robot's behavior, 285, 288
Roles, vii, 4, 7, 16, 18, 20, 21, 27, 45, 51–53, 56, 58–60, 67, 77, 79–81, 83, 93, 98, 102, 104, 105, 129, 138, 155, 157, 158, 160, 163, 164, 173, 175, 183, 192, 207, 208, 219, 222, 224, 246–256, 258, 263, 264, 269, 276, 277, 284, 286, 288, 290, 291
Roles of humans and AI, 248–256, 263, 264, 269

S
Science communication, 159, 171
Scientific collaboration (team, community, society), 155–184
Screenplay, 195, 196
Screenwriters, 198, 200–205, 209, 278
Script, 8, 52, 67, 195–205, 208, 209
Serendipity, 175–176, 178, 201, 202, 208, 209
Shared mental model, 130, 131, 144
Shared praxis, 175–179, 183
Social Process Assessment for Recognizing Creativity (SPARC), 127–149, 277
Social robot, viii, 245–270, 279, 281, 284–286, 289, 290
Societal trust in science, 184
Sociotechnical interaction, 290
Spatial, 32, 101, 200, 201, 207–209
Statistics Landscape Projections over Unseen Past (SLAPUP), 216, 220, 223–236, 240
Stats.Hakodate Browser, 216, 220, 223, 224, 233–241
Stereotype content model, 280, 291
Stoplight Model of Creativity, 131–133, 135
Story, 6, 8, 133, 134, 161, 166, 189–209, 246, 260, 261, 278, 286, 288, 289
Storytelling, 30, 189, 190, 209, 221, 251, 253, 287–289
Symmetry, 50–52, 63

T
Task, 1, 7–9, 12–14, 18, 21, 23, 25, 28, 30, 31, 48–52, 62, 63, 67, 76, 77, 80, 81, 83, 86, 87, 89, 90, 102, 130, 133, 143, 144, 161, 191, 246, 247, 250–254, 256, 258, 262, 263, 268, 269, 277, 279, 283, 285, 287–290, 292
  shared focus, 50, 67

Team, 4, 18–19, 47, 77, 89–90, 127–131, 155–184, 199, 258, 277
Team creativity, 32, 83, 107, 135
Techniques, 30, 62, 68, 77, 79, 84, 88, 89, 160, 167, 176, 240, 250–252, 262, 263, 267, 277
3D space, 193, 200, 203
Tools, viii, 20, 31, 32, 49, 51–54, 63, 67, 68, 90, 95, 107, 108, 131, 135, 147, 148, 160, 166, 167, 169, 172–175, 191, 198, 199, 202, 206, 208, 209, 218, 220, 228, 231, 242, 245–249, 251, 254–256, 258, 269, 278, 281, 284, 291

U
Uncanny valley, 283

V
Validation boundaries, 164–165, 176, 182
Values, 2–4, 10, 46, 49, 57–62, 67, 77, 81, 92, 94, 97, 101, 108, 138, 176, 180–183, 223, 225, 231–233, 240, 249, 250, 256
Virtual environment, 27–30
VR headset, 200

W
Wizard of Oz, 200

MIX
Papier aus verantwortungsvollen Quellen
Paper from responsible sources
FSC® C105338
FSC
www.fsc.org

If you have any concerns about our products,
you can contact us on
ProductSafety@springernature.com

In case Publisher is established outside the EU,
the EU authorized representative is:
Springer Nature Customer Service Center GmbH
Europaplatz 3, 69115 Heidelberg, Germany

Printed by Libri Plureos GmbH
in Hamburg, Germany